QUANTIFICATION IN SCIENCE

The VNR Dictionary of Engineering Units and Measures

QUANTIFICATION IN SCIENCE

The VNR Dictionary of Engineering Units and Measures

Michele Melaragno (Dr. C. E., P.E.)
Professor of Building Sciences
University of North Carolina at Charlotte

VNR VAN NOSTRAND REINHOLD
New York

Library of Congress Catalog Card Number 91-86
ISBN 0-442-00641-1

Manufactured in the United States of America

Published by Van Nostrand Reinhold
115 Fifth Avenue
New York, New York 10003

Chapman and Hall
2-6 Boundary Row
London, SE1 8HN

Thomas Nelson Australia
102 Dodds Street
South Melbourne 3205
Victoria, Australia

Nelson Canada
1120 Birchmount Road
Scarborough, Ontario M1K 5G4, Canada

16 15 14 13 12 11 10 9 8 7 6 5 4 3 2 1

Library of Congress Cataloging-in-Publication Data

Malaragno, Michele G.
Quantification in science : the VNR dictionary of engineering units and measures / Michele Melaragno.
p. cm.
ISBN 0-442-00641-1
1. Units—Handbooks, manuals, etc. 2. Weights and measures--Handbooks, manuals, etc. 3. Physics—Handbooks, manuals, etc. 4. Technology—Handbooks, manuals, etc. I. Title.
QC61.M34 1991
530′.0212—dc20 91-86
CIP

To the memory of my Mother and Father, and to the family on both sides of the ocean.

"Man is the measure of all things, of the existence of things that are, and of the nonexistence of things that are not."

Protagora's quotation in Plato's Socratic dialogues *Theaetetus*

Contents

Preface

In any scientific discipline, progress from the abstraction of concepts to factual quantification is unavoidable. Within the many fields of present-day science, quantification is a reality that touches daily a large percentage of the population, whether individuals are involved with sciences and technologies directly in their occupational endeavors, or in the scholastic learning processes, or in their roles as consumers of products as well as recipients of services.

With such a large spectrum of applications, the quantification of physical concepts requires a clear understanding of the fundamentals by a varied audience with different levels of education. For quite a few years I have felt the need for a publication that would explain the quantities used in the various sciences, especially in physics and technology, in simple terms, easily understood by a lay audience. I thought that such a work should include definitions of the units involved, their numerical values, the various systems in which they are incorporated, and a practical means of converting each unit into an equivalent one in a different system. Furthermore, to give a realistic context to the units, I thought that it would be beneficial to present them in the historical process in which they were conceived. Therefore, I have included biographical notes about those scientists whose names were used to name the units themselves. I further wished to relate such scientists to those who had the most prominent roles throughout history, with special regard to the scientists who emerged in the twentieth century. To this effect, the Nobel Prize winners in physics since the inception of the award have been introduced, in terms of their contributions to the progress of human knowledge. To show the great proliferation of modern sciences into many branches, I have presented an overall schematization of such branches, which I hope will orient the reader at the outset of this book.

MICHELE MELARAGNO
Charlotte, NC

Introduction

The need to measure goods by volume and weight to facilitate trading and the need to measure areas to divide land into parcels provided the prime incentive for establishing units of measure. These units eventually would serve as essential tools to facilitate the scientific process of investigation. The ancient world saw the spontaneous sprouting of an enormous variety of systems of units, which spread throughout the geographical areas inhabited by civilized populations. This enormous variety has extended not only geographically but also chronologically through history; and the only unifications that eventually occurred were those fostered by political needs, when centralized governments formed new local or national groupings. As the empirical investigation of the physical world began, and the need to quantify observed phenomena became more stringent, systems of units of measure proliferated and became more precise. Modern metrology has grown along with the physical sciences and thus is relatively young in comparison to the history of humanity. It was only in the nineteenth century that scientists succeeded in obtaining an almost universal system of units with the development of the metric system in France. The scientists of the world had long had a common literary medium of communication, the Latin language; but it took much longer for them to attain a scientific means of communication, represented by a universal system of units. Even more surprising has been the stubbornness of the English-speaking countries in standing apart and continuing to use medieval systems of units, in spite of the rest of the world. This schism, which was generated by the political separation in the nineteenth century between the British Empire and continental Europe, eventually will be resolved with the gradual acceptance of the metric system in the United Kingdom and the United States; but the need for conversion factors to pass from one system to another still exists. The use of conversion factors is further justified by the existence of many other systems, which are derived from the metric system or are part of those systems used in the United Kingdom or the United States.

The variety of the disciplines that have emerged from the expansion of twentieth century science is of such astonishing proportions that a great deal of confusion exists in the layperson's mind, which is daily reinforced by the voluminous mass of information received through the media. From physics to medicine, from engineering to biology, the world of science has enveloped most human beings in a variety of direct and indirect ways, making itself so relevant and essential that it is practically impossible to ignore it in its controlling role. Thus, it is important to disperse the fog of confusion that usually obscures the interrelationships of the many branches of scientific endeavor, which continue to proliferate. The layperson can definitely benefit from taking an overall look at the various scientific fields as they branch out, and can still have a sense of their cohesiveness in doing so.

The scientific development that began with the history of humankind and has evolved with the progress of civilization is not an abstraction, but is a concrete process activated by individuals who practically dedicated their lives to the continuous process of learning. Such persons through history have woven a continuous fabric of interconnected statements that constitute the main body of scientific knowledge. These individuals, with their unique contributions, have appeared historically at different times, generating a valuable continuum of scientific growth. To understand science and its evolution is to know who these people were and to remember at least the major events to which they contributed.

The languages in which science was written were many, for knowledge originated independently of geographical barriers. From

Assyrian to Egyptian, from Greek to Roman, a great variety of tongues eventually found some sort of unification through usage of the Latin language, which continued for centuries. However, a common language and a common system of units are still needed. With regard to the many systems of units still in existence, the scientists of the world continue to need clarity and consistent communications. Only thirty years ago, the International System of Units (SI) finally gained acceptance in the scientific communities of the world; but in engineering, for instance, the technical vocabulary still uses both metric and British systems in the daily routine of practical work.

Three concerns are addressed in this book: the need for clarification of the present-day state of science; the need to become acquainted with the human component of science, that is, with those persons who have contributed and are contributing to the accumulation of knowledge in the various scientific fields; and the need to clarify the vocabulary of science and get a general picture of the various physical quantities and systems of units now in use. Integrating these themes into a cohesive, coherent whole, this text is intended to serve as a practical reference book that will appeal to various groups of readers who will find it useful in its entirety as well as in the detailed informations that it offers.

QUANTIFICATION IN SCIENCE

The VNR Dictionary of Engineering Units and Measures

1
Schematic Organization of Modern Sciences

The evolution of the sciences is integrally linked to the evolution of civilization. A typical subdivision of this process usually includes the following historical periods: Babylonian, Greek, Roman, Medieval Chinese, Western Medieval, Beginning of Modern Science in the West, Nineteenth Century, and Contemporary. In this long time span, the rate of growth that science has attained in the twentieth century has no precedent. The stunning proliferation of the various branches of science that are presently in existence has created a large spectrum of disciplines that require some organization to provide an overview. Although the relationships between the branches of science may be organized in different ways, an attempt is made here to group the various branches into a coherent scheme. Although exhaustive and comprehensive, the following organization, which is derived from several sources, is just an example which could be altered and further developed.

A first general subdivision of modern science could include the following:

1. Mathematics
2. Physical Sciences

3. Earth Sciences
4. Biological Sciences
5. Technological Sciences
6. Medicine and Affiliated Disciplines
7. Social Sciences and Psychology

Each of these branches is further explored in the schematic subdivisions that follow:

1. Mathematics
    - Set Theory
    - Algebra
        - Arithmetic
        - Elementary
        - Multivariate
        - Linear
        - Multilinear
        - Structures
        - Group theory
        - Ring theory
    - Geometry
        - Euclidean
        - Non-Euclidean
        - Projective
        - Analytic
        - Trigonometric
        - Combinatorial
        - Differential
        - Algebraic
    - Analysis
        - Real
        - Complex
        - Differential equations
        - Functional
        - Fourier
        - Probability
        - Vector
        - Tensor

- Combinatorics
- Number Theory
  - Elementary
  - Algebraic
  - Analytic
  - Probabilistic
- Typology
  - General
  - Groups
  - Differential
  - Algebraic

2. Physical Sciences
   - Physics
     - Mechanics
     - Thermodynamics
     - Heat
     - Electricity
     - Magnetism
     - Sound
     - Optics
     - Quantum mechanics
     - States of matter
     - Nuclear and atomic physics
   - Interdisciplinary Fields
     - Astrophysics
     - Biophysics
     - Geophysics
   - Astronomy
     - Planetary and lunar sciences
     - Meteoritics
     - The study of comets, minor planets, the origin of the solar system
     - Astrophysics (the study of stars, galaxies, and the universe; cosmology and cosmogony)
   - Chemistry
     - Inorganic
     - Organic
     - Analytical
     - Physical

- Interdisciplinary Fields of Chemistry
  - Biochemistry
  - Geochemistry
  - Chemical engineering

3. Earth Sciences
   - Geological Science
     - Mineralogy
     - Petrology
     - Economic geology
     - Geochemistry
     - Geodesy
     - Geophysics
     - Structural geology
     - Volcanology
     - Geomorphology
     - Glacial geology
     - Geology (engineering, environmental, urban)
     - Historical geology
     - Paleontology
     - Stratigraphy
     - Sedimentology
     - Astrogeology
   - Hydrologic Sciences
     - Hydrology
     - Limnology
     - Glaciology
     - Oceanography
   - Atmospheric Sciences
     - Meteorology (turbulence, chemistry, analysis, dynamics, radiation, thermodynamics, cloud physics)
     - Climatology
     - Aeronomy (the study of the atmospheres of other planets)
4. Biological Sciences
   - Molecular Biology
     - Biochemistry
     - Biophysics
     - Genetics

- Cell Biology
  - Cancer research
  - Microbiology
  - Radiation biology
  - Tissue culture
  - Transplantation biology
- Organismic Biology
  - Botany
  - Ecology
  - Embryology
  - Ethology
  - Eugenics
  - Genetics
  - Gnotobiology
  - Morphology
  - Paleontology
  - Physiology
  - Zoology
- Population Biology
  - Biogeography
  - Comparative psychology
  - Ecology
  - Population genetics
  - Taxonomy

5. Technological Sciences
   - Engineering
     - Civil engineering
     - Aeronautical engineering
     - Chemical engineering
     - Electrical and electronics engineering
     - Mechanical engineering
     - Optical engineering
   - Agriculture
     - Soil science
     - Plant production
     - Animal production
     - Agricultural economics and management
     - Agricultural engineering

- Interdisciplinary Fields
    - Bionics
    - Systems engineering and operations research
    - Cybernetics, control theory, and information science

6. Medicine and Affiliated Disciplines
    - Hospital Residence Specialties
        - Radiology
        - Surgery
        - Obstetrics and gynecology
        - Urology
        - Ophthalmology and otolaryngology
        - Neurology
        - Psychiatry
        - Anesthesiology
        - Pathology
    - Other Clinical Specialties
        - Aerospace medicine
        - Medical jurisprudence
        - Occupational medicine
        - Public health
        - Endocrinology
        - Immunology
        - Toxicology
        - Tropical medicine
    - Nonclinical Specialties and the Basic Medical Sciences
        - Medical physiology
        - Pathological physiology
        - Nutrition
        - Pharmacology
        - Experimental therapeutics
        - Gerontology
    - Ancillary Medical Disciplines
        - Cytotechnology
        - Medical records
        - Medical technology
        - X-ray technology
7. The Social Sciences and Psychology
    - Anthropology (cultural and physical)

- Sociology
  - Criminology
  - Penology
  - Social psychology
  - Demography
  - Human geography
- Economics
  - Mathematical economics
  - Econometrics
  - Accounting
- Political Science
  - The study of public opinion
  - Public law
  - Public administration
  - Political systems
  - International relations
- Psychology
  - Physiological psychology
  - Social psychology

2
Scientists in Physics

As an anthropomorphic tendency led humans to create an Olympus of human gods to give a visual representation of an abstract theology, it is logical also to insert in the body of scientific knowledge the human characteristics of those scientists who contributed to its existence. This chapter lists some of the most prominent physicists, whose work generally has included major contributions to the physical sciences.

Physics as we know it today stems from the early beginnings of the Western civilization that had its cradle in Ancient Greece; so the list of scientists who created the body of knowledge starts approximately five centuries before the birth of Christ. The classification of modern sciences (see Chapter 1) shows the great progress that has been made since the unstructured early beginning of the scientific process, when theoretical and empirical analyses of inductive and deductive explanations of the universe were holistically explored. The distinction of physics per se from other areas of knowledge promoted in the early schools of Ancient Greece was not clearly possible. Philosophy, physics, and mathematics, all structured on logic, were so strongly interwoven that it would be hard to label the practitioners, or to clearly separate philosophers from physicists and mathematicians. Therefore, it is not a simple task to select those scientists who could strictly be considered physicists in such early times, and some arbitrary distinctions have been made to limit the number of those individuals who were prominently associated with inductive processes. For instance, omitting Aristotle, Plato, and Pythagoras from the list may seem arbitrary, but it is necessary by the criteria used here.

Because the time period considered is so long, the value of this list of scientists is mostly historical, and the selection of the scientists mentioned herein is subjective. The exclusion of many excellent contributors to scientific progress is hard to justify, but because of space limitations a cutoff point had to be established. For the twentieth century, when the rate of scientific growth has been extraordinary, the selection was particularly difficult; so it was decided that Chapter 3 would present only those scientists who had been recipients of the Nobel Prize in physics, from the institution of the award to the most recent recipients.

Abbe, Ernst (1840–1905). Born in Eisenach, Thuringia (Germany). Physicist. Optics.

Alfven, Hannes Olof Gosta (1908–). Born in Norrkoping, Sweden. Astrophysicist. Magnetohydrodynamics.

Alhazen (*ca* 965–1038). Born in Basra, Iraq. Scientist. Theory and optics.

Alter, David (1907–1981). Born in Westmoreland County, Pennsylvania, USA. Inventor; physicist. Spectroscopy.

Ampere, Andre-Marie (1775–1836). Born in Polemieux, France. Physicist, mathematician, chemist, and philosopher. Electromagnetics.

Anderson, Carl David (1905–). Born in New York City, USA. Physicist. Particle physics.

Anderson, Philip Warren (1923–). Born in Indianapolis, Indiana, USA. Physicist. Solid state physics.

Ångstrom, Anders Jonas (1814–1874). Born in Logdo, Sweden. Physicist and astronomer. Spectroscopy.

Appleton, Edward Victor (1892–1965). Born in Bradford, Yorkshire, England. Physicist. Radio waves.

Arago (Dominique) François (1786–1853). Born in Estagel, France. Scientist. Physics and astronomy.

Archimedes (*ca* 287–212 B.C.). Born in Syracuse, Sicily. Mathematician and physicist. Statics and hydrostatics.

Armstrong, Edwin Howard (1890–1954). Born in New York City, USA. Electronics engineer. Radio.

Bacon, Roger (*ca* 1220–1292). Born in Bisley, Gloucestershire, England. Philosopher and scientist. Experimentation and conclusions.

Bainbridge, Kenneth Tompkins (1904–). Born in Cooperstown, New York, USA. Physicist. Mass spectrometer.

Balmer, Johann Jakob (1825–1898). Born in Lausanne, Switzerland. Mathematical reactor. Formulae of the frequencies of atomic spectral lines.

Bardeen, John (1908–). Born in Madison, Wisconsin, USA. Physicist. Transistor; superconductivity.

Barkla, Charles Glover (1877–1944). Born in Widnes, Lancashire, England. Physicist. X-rays and ionizing radiation.

Becquerel, Antoine-Henri (1852–1908). Born in Paris, France. Physicist. Radioactivity.

Bernoulli, Daniel (1700–1782). Born in Graningen, Holland. Physicist and mathematician. Hydrodynamics.

Bethe, Hans Albrecht (1906–). Born in Strasbourg, Germany. American physicist. Energy production in stars.

Bhabha, Homi Jehangir (1909–1966). Born in Bombay, India. Theoretical physicist. Behavior of subatomic particles.

Black, Joseph (1728–1799). Born in Bordeaux, France. Scottish physicist and chemist. Thermodynamics.

Blackett, Lord Patrick Maynard Stuart (1897–1974). Born in Croydan, Surrey, England. Physicist. Atomic transmulation and nuclear reactions.

Boltzmann, Ludwig (1866–1906). Born in Vienna, Austria. Theoretical physicist. Kinetic theory of gases, electromagnetism, and thermodynamics.

Born, Max (1882–1970). Born in Breslau, Germany. British physicist. Quantum mechanics.

Bose, Satyendranath (1894–1974). Born in Calcutta, India. Physicist and mathematician. Nuclear physics; statistics.

Bowden, Frank Philip (1903–1968). Born in Hobart, Tasmania, Australia. Physicist and chemist. Electrochemistry.

Boys, Charles Vernon (1855–1944). Born in Wing, Rutland, England. Inventor and physicist. Scientific apparatus.

Bragg, William Henry (1862–1962), and *Bragg (William) Lawrence* (1890–1971). Born in Westward, Cumberland, England, and Adelaide, South Australia. Physicists. X-ray diffraction.

Branley, Edouard Eugene Desire (1866–1940). Born in Amiens, France. Physicist. Wireless telegraphy and radio.

Braun, Karl Ferdinand (1850–1918). Born in Fulda, Germany. Physicist. Wireless telegraphy.

Brewster, David (1781–1868). Born in Jedburgh, Scotland. Physicist. Polarization of light; kaleidoscope.

Bridgman, Percy Williams (1882–1961). Born in Cambridge, Massachusetts, USA. Physicist. Behavior of materials at high temperature and pressure.

Bullard, Edward Crisp (1907–1980). Born in Norwich, England. Geophysicist. Marine geophysics.

Cailletet, Louis Paul (1832–1913). Born in Chatillon-sur-Seine, France. Physicist and inventor. Liquefaction of the permanent gases.

Carnot, Nicholas Leonard Sadi (1796–1832). Born in Paris, France. Physicist. Thermodynamics.

Cavendish, Henry (1731–1810). Born in Nice, France. British physicist and chemist. Gravitational constant.

Chadwick, James (1891–1976). Born in Bollington, Cheshire, England. Physicist, Neutron.

Charles, Jacques Alexancer Cesar (1746–1823). Born in Beaugency, Loiret, France. Physicist and mathematician. Expansion of gases.

Chladni, Ernst Florens Friedrich (1756–1827). Born in Wittenberg, Saxony, Germany. Physicist. Acoustics.

Clausius, Rudolf Julius Emmanuel (1822–1888). Born in Koshin, Poland. German theoretical physicist. Thermodynamics.

Cockcroft, John Douglas (1897–1967). Born in Todmorden, Yorkshire, England. Physicist. Particle accelerator; artificial nuclear transformation.

Compton, Arthur Holly (1892–1962). Born in Wooster, Ohio, USA. Physicist. Compton effect.

Coriolis, Gaspard Gustave de (1792–1843). Born in Paris, France. Physicist. Coriolis force.

Coulomb, Charles (1736–1806). Born in Angouleme, France. Physicist. Electric charge and magnetism.

Crookes, William (1832–1919). Born in London, England. Physicist and chemist. High-voltage discharge tubes.

Daniell, John Frederic (1790–1845). Born in London, England. Meteorologist, inventor, and chemist. Daniell cell (electricity).

Davisson, Clinton Joseph (1881–1958). Born in Bloomington, Illinois, USA. Physicist. Wave nature of electrons.

Democritus (*ca* 460–370 B.C.). Born in Abdera, Thrace. Greek philosopher. Atomic theory of matter.

Desormes, Charles Bernard (1777–1862). Born in Dijon, Côte d'Or, France. Physicist and chemist. Ratio of the specific heats of gases.

Dewar, James (1862–1923). Born in Kincardine-on-Forth, Scotland. Physicist and chemist. Cryogenics.

Dicke, Robert Henry (1916–). Born in St. Louis, Missouri, USA. Physicist. Cosmology.

Dirac, Paul Adrien Maurice (1902–1984). Born in Bristol, England. Theoretical physicist. Quantum electrodynamics.

Doppler, (Johann) Christian (1803–1853). Born in Salzburg, Austria. Physicist. Doppler effect (frequency of waves).

Einstein, Albert (1879–1955). Born in Ulm, Germany. American theoretical physicist. Theories of relativity.

Fabry, Charles (1867–1965). Born in Marsaille, France. Physicist. Optics.

Fahrenheit, Daniel Gabriel (1686–1736). Born in Danzig, Poland. Dutch physicist. Thermometers and Fahrenheit scale of temperature.

Faraday, Michael (1791–1867). Born in Newingham, Surrey, England. Physicist and chemist. Electricity.

Fermi, Enrico (1901–1954). Born in Rome, Italy. American physicist. Development of atomic bomb.

Fitch, Val Lodgson (1923–), and *Cronin, James Watson* (1931–). *Fitch* was born in Merriman, Nebraska, USA; *Cronin* in Chicago, Illinois, USA. Physicists. Particle physics.

Fitzgerald, George Francis (1851–1901). Born in Dublin, Ireland. Theoretical physicist. Electromagnetic theory of light and radio waves.

Fizeau, Armand Hippolyte Louis (1819–1896). Born in Paris, France. Physicist. Speed of light on the earth's surface.

Fortin, Jean Nicholas (1750–1831). Born in Mouchy-la-Ville, France. Instrument maker. Mercury barometer.

Foucault, Jean Bernard Leon (1819–1868). Born in Paris, France. Physicist, gyroscope, rotation of the earth, velocity of light.

Franck, James (1882–1966). Born in Hamburg, Germany. American physicist. Quantum theory of Max Plank.

Franklin, Benjamin (1706–1790). Born in Boston, Massachusetts, USA. Scientist. Electrical positive and negative charges.

Fraunhofer, Joseph von (1787–1826). Born in Strubing, Germany. Physicist and optician. Spectroscope.

Fresnel, Augustin Jean (1788–1827). Born in Broglie, Normandy. French physicist. Transverse-wave theory of light.

Frisch, Otto Robert (1906–1979). Born in Vienna, Austria. British physicist. Atomic fission.

Gabor, Dennis (1900–1979). Born in Budapest, Hungary. British physicist. Holography.

Galileo (1564–1643). Born in Pisa, Italy. Physicist and astronomer. Laws of motion of falling bodies.

Galvani, Luigi (1737–1798). Born in Bologna, Italy. Anatomist. Electric currents.

Gamow, George (1904–1968). Born in Odessa, Russia. American physicist. Theory of origin of the universe.

Gauss, Carl Friedrich (1777–1844). Born in Brunswick, Germany. Mathematician and physicist. Terrestrial magnetism.

Geiger, Hans Wilhelm (1882–1945). Born in Neustadt, Rheinland, Pfalz, Germany. Physicist. Detecting radioactivity.

Gell-Mann, Murray (1929–). Born in New York City, USA. Theoretical physicist. Subatomic particles.

Gilbert, William (1544–1603). Born in Colchester, Essex, England. Physician and physicist. Magnetism.

Giorgi, Giovanni (1871–1950). Born in Lucca, Italy. Civil engineer and professor at the University of Rome. Invention of the (MKSA) system of units.

Glaster, Donald Arthur (1926–). Born in Cleveland, Ohio, USA. Physicist. Bubble chamber.

Goldstein, Eugene (1850–1930). Born in Gleiwitz, Poland. German physicist. Electrical discharges through gases at low pressures.

Grimaldi, Francesco Maria (1618–1663). Born in Bologna, Italy. Physicist. Diffraction of light.

Hahn, Otto (1879–1968). Born in Frankfurt, Germany. Radiochemist. Nuclear fission.

Harrison, John (1693–1776). Born in Foulky, Yorkshire, England. Instrument maker. Chronometers.

Hawking, Stephen William (1942–). Born in Oxford, England. Physicist. Cosmology.

Heaviside, Oliver (1850–1925). Born in Camden Town, London, England. Physicist and electrical engineer. Passage of electrical waves through the atmosphere.

Heisenberg, Werner Karl (1901–1976). Born in Duisberg, Germany. Physicist. Quantum mechanics and the uncertainty principle.

Helmholtz, Hermann Ludwig Ferdinand von (1821–1894). Born in Potsdam, Germany. Physicist and physiologist. Conservation of energy.

Henry, Joseph (1797–1878). Born in Albany, New York, USA. Physicist. Electromagnetic induction.

Hertz, Heinrich Rudolf (1857–1894). Born in Hamburg, Germany. Physicist. Radio waves.

Herzberg, Gerhard (1906–). Born in Hamburg, Germany. Canadian physicist. Electronic structure and geometry of molecules.

Hess, Victor Francis (1883–1964). Born in Waldstein, Austria. American physicist. Cosmic rays.

Hooke, Robert (1635–1703). Born in Freshwater Isle of Wight, England. Physicist. Derivation of Hooke's law of elasticity.

Huygens, Christiaan (1629–1695). Born in The Hague, Netherlands. Physicist and astronomer. Pendulum.

Jensen, Johannes Hans Daniel (1907–). Born in Hamburg, Germany. Physicist. Atomic nuclei.

Josephson, Brian David (1960–). Born in Cardiff, England. Physicist. Tunneling effect in superconductivity.

Joule, James Prescott (1818–1889). Born in Salford, England. Physicist. Conservation of energy.

Kapitza, Pyotr Leonidovich (1894–). Born in Kronstadt, Russia. Physicist. Superfluidity of liquid helium.

Kelvin, Lord (1826–1907). Born in Belfast, Ireland. British physicist. Absolute scale of temperature.

Kennelly, Arthur Edwin (1861–1939). Born in Colaba, Bombay, India. American physicist. Predicting the existence of the ionosphere.

Kerr, John (1824–1907). Born in Ardrossan, Ayrshire, England. Physicist. Magnetism and electricity.

Kirchhoff, Gustav Robert (1824–1887). Born in Königsberg, Germany. Physicist. Science of spectroscopy.

Kundt, August Adolph (1839–1894). Born in Schwerin, Mecklenburg, Germany. Physicist. Velocity of sound in gases and solids.

Lande, Alfred (1888–1975). Born in Elberfeld, Germany. American physicist. Splitting factor in quantum theory.

Langevin, Paul (1872–1946). Born in Paris, France. Physicist. Generating ultrasonic waves.

Laue, Max Theodor Felix von (1879–1960). Born in Pfaffendorf, Germany. Physicist. X-rays.

Lawrence, Ernest Orlando (1901–1958). Born in Canton, South Dakota, USA. Physicist. Cyclotron.

Lebedev, Pyotr Nikolayevich (1866–1912). Born in Moscow, Russia. Physicist. Pressure that light exerts on bodies.

Leclanche, Georges (1839–1882). Born in Paris, France. Engineer. Battery or dry cell.

Lenard, Philipp Edward Anton (1862–1947). Born in Pozsony, Hungary. German physicist. Photoelectric effect.

Lenz, Heinrich Friedrich Emil (1804–1865). Born in Dorpat, Russia. Physicist. Laws of electromagnetism.

Lindemann, Frederick Alexander (1886–1957). Born in Baden-Baden, Germany. British physicist. Advancement of the quantum theory.

Lissajous, Jules Antoine (1822–1880). Born in Versailles, France. Physicist. Wave motion.

Lizhi, Fang (1936–). Born in Peking, China. Astrophysicist. Cosmology.

Lodge, Oliver Joseph (1851–1940). Born in Penkhull, Strafford-shire, England. Physicist. Radio.

Lorentz, Hendrick Antoon (1853–1928). Born in Arnhem, Holland. Physicist. Theory of electromagnetism.

Lorenz, Ludwig Valentin (1829–1891). Born in Elsimze, Denmark. Mathematician and physicist. Heat, electricity, and optics.

Lummer, Otto Richard (1860–1925). Born in Jena, Saxony, Germany. Physicist. Optics.

Lyman, Theodore (1874–1954). Born in Boston, Massachusetts, USA. Physicist. Spectroscopics in ultraviolet region.

Mach, Ernst (1838–1916). Born in Chirlitz-Turas, Austria. Physicist. Velocities.

Maiman, Theodore Harold (1927–). Born in Los Angeles, California, USA. Physicist. Laser.

Malus, Etienne Louis (1775–1812). Born in Paris, France. Physicist. Polarized light.

Maxwell, James Clerk (1831–1879). Born in Edinburgh, Scotland. Physicist. Light.

Mayer, Julius Robert (1814–1878). Born in Heilbronn, Germany. Physicist. Conservation of energy.

Meitner, Lise (1878–1968). Born in Vienna, Austria. Swedish physicist. Radioactive decay.

Michelson, Albert Abraham (1852–1931). Born in Strelno, Germany. American physicist. Light.

Millikan, Robert Andrews (1868–1953). Born in Morrison, Illinois, USA. Physicist. Electrons.

Morley, Edward Williams (1838–1923). Born in Newark, New Jersey, USA. Physicist and chemist. Light.

Moseley, Henry Gwyn Jeffreys (1887–1915). Born in Weymouth, England. Physicist. Atom.

Mossbauer, Rudolf Ludwig (1929–). Born in Munich, Germany. Physicist. Radiation of an atomic nucleus.

Mott, Nevill Francis (1905–). Born in Leeds, Great Britain. Physicist. Semiconductors.

Newton, Isaac (1642–1727). Born in Woolsthorpe, Lincolnshire, England. Physicist and mathematician. Laws of motion.

Nicol, William (1768–1851). Born in Scotland. Physicist and geologist. Light.

Nobili, Leopoldo (1784–1835). Born in Trassilico, Italy. Physicist. Electrochemistry and thermoelectricity.

Oersted, Hans Christian (1777–1851). Born in Rudkbing, Langeland, Denmark. Physicist. Electromagnetism.

Ohm, Georg Simon (1789–1954). Born in Erlangen, Bavaria, Germany. Physicist. Electrical resistance.

Onnes, Heike Kamerlingh (1853–1926). Born in Groningen, Denmark. Physicist. Properties of matter at low temperature.

Oppenheimer, Julius Robert (1904–1967). Born in New York, USA. Physicist. Quantum mechanics.

Pascal, Blaise (1623–1662). Born in Clermont-Ferrand, France. Mathematician and physicist. Pressure, hydraulics.

Pauli, Wolfgang (1900–1958). Born in Vienna, Austria. Swiss physicist. Quantum theory.

Peierls, Rudolf Ernst (1907–). Born in Berlin, Germany. British physicist. Quantum theory and nuclear physics.

Peregrinus, Petrus (*ca* 1220). Born in France. Scientist and scholar. Magnetism.

Perrin, Jean Baptiste (1870–1942). Born in Lille, France. Physicist. Atoms.

Pippard, (Alfred) Brian (1920–). Born in London, England. Physicist. Superconductivity.

Pixii, Hippolyte (1808–1835). Born in France. Inventor. Electricity generator.

Planck, Max Karl Ernst Ludwig (1858–1947). Born in Kiel, Germany. Physicist. Energy; quantum theory.

Plucker, Julius (1801–1868). Born in Elberfeld, Germany. Mathematician and physicist.

Poisson, Simeon Denis (1781–1840). Born in Pithiviers, Loizet, France. Mathematician and physicist. Elasticity of material.

Powell, John Henry (1852–1914). Born in Monton, Lancashire, England. Physicist, mathematician, and inventor. Electromagnetic energy.

Prandtl, Ludwig (1875–1953). Born in Freising, Germany. Physicist. Fluid mechanics, aerodynamics.

Prevost, Pierre (1751–1839). Born in Geneva, Switzerland. Physicist. Heat radiation from bodies.

Pringsheim, Ernst (1859–1917). Born in Breslau, Germany. Physicist. Thermal radiation.

Rainwater, (Leo) James (1917–). Born in USA. Physicist. Structure of the atomic nucleus.

Raman, Chandrasekhara Venkata (1888–1970). Born in Trichinopoly, Madras, India. Physicist. Light.

Rayleigh, Lord (1842–1919). Born in Langford Grove, Essex, England. Physicist. Classical physics.

Ritter, Johann Wilhelm (1776–1810). Born in Samnetz, Silesia, Poland. German physicist. Electrocytic cells and ultraviolet radiation.

Röntgen, Wilhelm Konrad (1845–1923). Born in Lennex, Prussia. Physicist. X rays.

Rowland, Henry Augustus (1848–1901). Born in Honesdale, Pennsylvania, USA. Physicist. Analysis of spectra.

Rumford, Count Benjamin Thompson (1753–1814). Born in Woburn, Massachusetts, USA. Physicist. Heat as form of motion.

Rutherford, Ernest (1871–1937). Born in Nelson, New Zealand. British physicist. Nuclear physics (radioactivity).

Rydberg, Johannes Robert (1854–1919). Born in Halmstad, Sweden. Physicist. Mathematical expression of frequencies.

Sabine, Edward (1788–1883). Born in Dublin, Ireland. Physicist. Terrestial magnetism.

Sakharov, Andrei Dmitriyevich (1921–). Born in Moscow, Russia. Physicist. Thermonuclear weapons.

Schrödinger, Erwin (1887–1961). Born in Vienna, Austria. Physicist. Mechanics; atomic structure.

Shaw, William Napier (1854–1965). Born in Birmingham, England. Meteorologist. Atmospheric pressure.

Simon, Franz Eugen (1893–1956). Born in Berlin, Germany. British physicist. Third Law of Thermodynamics.

Simpson, George Clark (1878–1965). Born in Derby, England. Meteorologist. Atmospheric electricity.

Snell, Willebord (1580–1626). Born in Leiden, Poland. Physicist. Law of refraction.

Sommerfeld, Arnold (1868–1951). Born in Königsberg, Prussia. German physicist. Quantum theory of atomic structure.

Stark, Johannes (1876–1957). Born in Schickenhef, Bavaria, Germany. Physicist. Electric discharge.

Stefan, Josef (1835–1893). Born in Klagenfurt, Austria. Physicist. Radiation of bodies.

Stern, Otto (1888–1969). Born in Sohrau, Upper Silesia, Germany. American physicist. Atoms and molecules.

Stevinus, Simon (*ca* 1548–1620). Born in Bruges, Belgium. Scientist. Statics and hydrodynamics.

Stokes, George Gabriel (1819–1903). Born in Skreen, Sligo, Ireland. Physicist. Fluids.

Stoney, Geroge Johnstone (1826–1911). Born in Oakley Park, King's County, Ireland. Physicist. Electrons.

Sutherland, Gordon (1907–1980). Born in Caithness, Scotland. Physicist. Infrared spectroscopy.

Tabor, David (1913–). Born in London, England. Physicist. Tribology.

Tesla, Nikola (1856–1943). Born in Smiljan, Croatia, Yugoslavia. American physicist. Alternating current electricity.

Thomson, George Paget (1892–1975). Born in Cambridge, England. Physicist. Electron diffraction.

Thomson, James (1822–1892). Born in Belfast, England. Physicist and engineer. Hydrodynamics.

Thomason, Joseph John (1856–1940). Born in Cheetham Hill, England. Physicist. Electrons; conduction of electricity through gases.

Tolansky, Samuel (1907–1973). Born in Newcastle-upon-Tyne, England. Physicist. Spectroscopy and interferometry.

Torricelli, Evangelista (1608–1647). Born in Faenza, Italy. Physicist and mathematician. Invention of the barometer.

Townes, Charles Hard (1915–). Born in Greenville, South Carolina, USA. Physicist. Theory of the maser.

Townsend, John Sealy Edward (1868–1957). Born in Galway, Ireland. Mathematical physicist. Kinetics of electrons and ions in gases.

Tyndall, John (1820–1893). Born at Leighlinbridge, Conlow, Ireland. Physicist. Light.

Van Allen, James Alred (1914–). Born in Mount Pleasant, Iowa, USA. Physicist. U.S. space program.

Van de Graaff, Robert Jemison (1901–1967). Born in Tuscaloosa, Alabama, USA. Physicist. Built the electrostatic high voltage generator.

Van Vleck, John Hasbrouck (1899–1980). Born in Middletown, Connecticut, USA. Physicist. Magnetism.

Vening Meinesz, Felix Andries (1887–1966). Born in The Hague, The Netherlands. Geophysicist. Geophysics and geodesy.

Volta, Alessandro (1745–1827). Born in Como, Italy. Physicist. Electric current and electric battery.

Von Gueicke, Otto (1602–1686). Born in Magdeburg, Germany. Physicist. Invented the air pump and static electricity generator.

Von Neumann, Johann (1903–1957). Born in Budapest, Hungary. American physicist and mathematician. Concepts of programming computers.

Walton, Ernest Thomas Sinton (1903–). Born in Dungarvan, Ireland. Physicist. Particle accelerator and artificial transmutation.

Waterson, John James (1811–1883). Born in Edinburgh, Scotland. Physicist. Kinetic theory of gases.

Weber, Wilhelm Eduard (1804–1891). Born in Wittenberg, Germany. Physicist. Electricity and magnetism.

Wheatstone, Charles (1802–1875). Born in Gloucester, England. Physicist. Electrical resistance.

Wheeler, John Archibald (1911–). Born in Jacksonville, Florida, USA. Physicist. Atomic and nuclear physics.

Wien, Wilhelm (1844–1928). Born in Gaffken, East Prussia. German physicist. Thermal radiation.

Wilson, Charles Thomson Rees (1869–1959). Born near Glencorse, Scotland. Physicist. Atomic particles detector.

Young, Thomas (1773–1829). Born in Milverton, Somerset, England. Physicist and physician. Light and physiology of vision.

Yukawa, Hideki (1907–). Born in Kyoto, Japan. Physicist. Elementary particles and nuclear forces.

Zeeman, Pieter (1865–1943). Born in Zonnemaire, Zeeland, Netherlands. Structure of the atom.

3
Nobel Prize Winners in Physics

To select scientists who have achieved particular distinction in the development of modern physics, one can do no better than to follow the judgment of the Royal Swedish Academy of Sciences, which annually presents the world's most prestigious science awards. The lists of physicists who have received the Nobel Prize from its inception in 1901 through 1989 is included hereafter, with concise statements about the primary field of endeavor in which each scientist worked as well as the motivation for the award.

The awarding of the Nobel Prizes was instituted by Alfred Bernhard Nobel (1833–1896), the inventor of dynamite (TNT), which had brought him great international fame and affluence. With the establishment of the extraordinarily prestigious prizes, Nobel succeeded in being remembered for something more humanitarian than dynamite, with its deadly potential when used in the context of warfare.

The prizes are limited to the following five categories: chemistry, physics, physiology or medicine, literature, and contributions for world peace, excluding other fields in order to avoid an inflated and too disperse distribution of awards.

The following brief descriptions of the accomplishments of the various Nobel Prize winners in physics will give the reader a general sense of the directions in which modern physics has been developing. Furthermore, from the brief biographical data given here, it is possible to observe the geographical distribution of those areas where the physical sciences have progressed the most.*

Alfvén, Hannes. Born: May 30, 1908, Norrkoping, Sweden. Specialized in plasma physics, and is best known for the identification of magnetohydrodynamical (MHD) waves (Alfven waves) used in astrophysical and nuclear fusion problems. Recipient of the Nobel Prize in Physics for 1970, along with Louis Néel of France, for explaining the forces acting in astrophysics, as in the sun's origin, the formation of the planetary system, the exchange of energy rotation from the sun to the planets, supernovae, and the eruptions coming from the center of the galaxy.

Alvarez, Luis W. Born: June 13, 1911, San Francisco, California; died: September 1, 1988, Berkeley, California. Specialized in high-energy particle physics, and is best known for making the hydrogen bubble chamber into a precise instrument. Recipient of the Nobel Prize in Physics for 1968 for perfecting a method for tracking elementary particles.

Anderson, Carl David. Born: September 3, 1905, New York, New York. Specialized in particle physics, and is best known for the discovery of the positron during his investigations of cosmic radiations. Recipient of the Nobel Prize for Physics for 1936, along with Victor Franz Hess, for the discovery of the positron. Note that the positron (the antiparticle of the electron) was created from a collision between cosmic-ray particles and air molecules.

Anderson, Philip W. Born: December 13, 1923, Indianapolis, Indiana. Specialized in solid-state physics, and is best known for electrical and magnetic properties of solid materials and for modeling

---

**The highest concentration of Nobel Prize winners in physics has occurred in the United States with 34 American-born winners in a total of 136 winners worldwide, not counting those persons who moved to the United States from other countries.*

the mutual interactions of electrons and their motions in materials lacking in crystalline structure. Recipient of the Nobel Prize in Physics for 1977, along with Sir Nevill Mott and John H. Van Vleck, for recognizing large-scale regularities in the highly disordered motions of electrons.

Appleton, Sir Edward Victor. Born: September 6, 1892, Bradford, England; died: April 21, 1965, Edinburgh, Scotland. Specialized in radio and atmospheric physics, and is best known for locating the Heaviside layer (from Oliver Heaviside) at 100 kilometers over the earth and for identifying the Appleton layer (named after him), which splits into two layers during the day and merges into one at night, located at an altitude of 230 kilometers. Recipient of the Nobel Prize in Physics for 1947 for the determination of radio wave frequencies affecting the interference of ground waves with reflected waves. Note that radio waves of various frequencies penetrate or are reflected from layers in proportion to their state of ionization, which is affected by various degrees of rarefication of the atmosphere and by the variability of the sunspots from year to year.

Bardeen, John. Born: May 23, 1908, Madison, Wisconsin. Specialized in solid-state physics, and is best known for the development of the transistor that replaced the vacuum tube. Recipient of the Nobel Prize in Physics twice: for 1956, along with William Shockley and Walter Houser Brattain, for their work in the area of semiconductors which eventually led to the discovery of transistors; and for 1972, along with Leon N. Cooper and John Robert Schrieffer, for their contribution to superconductivity. Notice that superconductivity is a phenomenon observed in metals as their temperature gets close to absolute zero, which consists of losing most of their resistance to the flow of electricity. Practical applications of this phenomenon include high-efficiency energy transmissions (in power lines) and the development of high-velocity trains running on superconductive tracks.

Barkla, Charles Glover. Born: June 7, 1877, Widnes, Lancashire, England; died: October 23, 1944, Edinburgh, Scotland. Specialized in X-radiation and secondary radiation, and is best known for his studies on the characteristics of elements, by demonstrating that the position of each element in the periodic table depends on the electric charge of the atom and that each element exhibits a different X-ray spectrum. Recipient of the Nobel Prize in Physics for 1918 for the discovery of the secondary X ray produced when a sample of an element is exposed to X rays. Note that such secondary X rays were observed to be of two types (K and L series); the K-series has almost the same characteristics as the primary X rays, including penetrability and can almost be considered to be a diffusion of the primary X rays, whereas the L-series instead is independent of the primary X rays and varies for each element.

Basov, Nikolay Gennadiyevich. Born: December 14, 1922, Usman, near Vornezh, Soviet Union. Specialized in quantum electronics, and is best known for his invention of quantum microwave amplification devices (masers) and light amplifiers (lasers) that depend on stimulated emissions of radiation, predicted by Einstein in theoretical terms. Recipient of the Nobel Prize in Physics for 1964, along with Aleksandr Mikhailovich Prokhorov and Charles Townes, for producing the first maser in the Soviet Union and for his contributions to semiconductors used in lasers. A typical example is demonstrated by the ruby laser. This is a powerful beam of monochromatic, coherent light (when the crests of the light waves coincide) that emerges from the ruby crystal. To produce it, a xenon flash lamp is used to create an inverted population in the ruby, in which a majority of chromium atoms are set in a high-energy state and a minority of them are put in a ground state (zero energy). Such atoms at the ground state generate photons that stimulate radiations. Such radiations, in turn, are reflected by two face-to-face mirrors that make the radiations cross the ruby numerous times. Under these conditions, the ruby laser is eventually obtained.

Becquerel, Antoine-Henri. Born: December 15, 1852, Paris, France; died August 25, 1908, Le Croisic, France. Specialized in radioactivity, and is best known for the discovery of beta and gamma rays, which are spontaneously generated by uranium. Discovered in 1896, these radiations—originally named "Bequerel rays" after this scientist—have the ability to make other bodies in their vicinity temporarily radioactive. A major consequence of his work was the subsequent discovery of other radioactive elements (thorium, polonium, radium, and actinium) by Pierre and Marie Curie. Recipient of the Nobel Prize in Physics for 1903, along with Pierre and Marie Curie, for the discovery of the radioactivity of uranium and for the identification of the "Bequerel rays." Bequerel's discoveries were in turn inspired by the work of Wilhelm Conrad Röntgen of Germany, winner of the 1895 Nobel Prize for the discovery of cathode rays which generate X rays.

Bednorz, J. Georg. Born: May 16, 1950, Neuenkirchen, West Germany. Specialized in solid-state physics and superconductivity, and is best known for discovering superconductive materials operating at a relatively high temperature, allowing a practical application of superconductivity principles that in turn has recently opened the horizon to other researchers around the world. For further clarification, consider that certain substances can transport large amounts of electricity with minimum resistance when cooled to extremely low temperatures; more precisely, at a certain critical temperature the electrical resistance is zero, and the phenomenon is called superconductivity. Solid mercury, for instance, attains zero electrical resistance when cooled down to 4 degrees Celsius above absolute zero, as was discovered by Heike Kamerlingh Onnes in 1911. Such low temperatures are not practical, and it was necessary to find materials that could superconduct at higher temperatures. Recipient of the Nobel Prize in Physics for 1987, along with Karl Alexander Muller, for discovering the property of superconductivity of a ceramic substance consisting of lanthanum, barium, copper, and oxygen, which superconducts at 35 degrees Celsius above absolute zero, a much higher temperature than that used for any other substances. Note that such a substance, originally produced by a French chemist, was eventually tested for superconductivity by Bednorz and Muller in the IBM Research Division of the Zurich Research Laboratory in Rorschach, Switzerland.

Bethe, Hans Albrecht. Born: July 2, 1906, Strasbourg, Germany. Specialized in nuclear physics and astrophysics, and is best known for explaining energy production in the sun and other stars, through nuclear reactions that occur at different temperatures. In the sun, at a temperature of 16 million degrees Celsius, he proposed that the possible nuclear reactions include carbon, nitrogen, and oxygen, whereas for stars having higher temperatures the expected nuclear reactions include carbon, helium, and hydrogen. In the latter type of reactions (carbon cycle), hydrogen is transformed into helium under the catalytic action of carbon, and it is through such nuclear reactions that great amounts of energy are released. Of major relevance is the conclusion that temperature is a controlling factor for the type of nuclear reactions that occur in stars, and, therefore, temperature is a major parameter influencing the life cycles of stars in the universe. Note that it was only with the advent of computers that Bethe was able to ascertain the exact temperature of the sun to be 16 million degrees Celsius, correcting his previous estimate of 19 million degrees Celsius and thus refining his theory. Recipient of the Nobel Prize in Physics for 1967 for the discovery of the carbon cycle and the influence of temperature on the type of nuclear reactions occurring in the sun and other stars.

Binnig, Gerd. Born: June 20, 1947, Frankfurt, West Germany. Specialized in scanning tunneling microscopy, and is best known for his studies in the development of high-technology instruments capable of probing the structure of matter within the scale of atomic and subatomic particles. Such instrumentation implied new theoretical concepts that transcended the conventional visual exploration of form using light. In other words, when the dimensions of the structures to be explored are smaller than the wavelength of light, optical instrumentation is no longer possible, and the system of exploring forms has to be based on another medium. The ultramicroscope, which is still based on light as the medium of exploration, cannot be used for atomic and subatomic observations. Recipient of the Nobel Prize in Physics for 1986, along with Heinrich Rohrer and Ernst Ruska, for the development of scanning tunneling microscopy. He and Rohrer had worked together on such instrumentation in the same laboratory in Zurich. Their work capitalized on Ernst Ruska's previous discovery of the electron microscope, developed in 1930, adding substantial refinements 56 years later. Note that in this process a very fine needle with a sensitive tip, so fine as to reach atomic dimensions, can explore surfaces with an accuracy on the atomic scale. To maintain such accuracy of measurement, the instrumentation includes the interposition of an electric layer between the needle and the surface, avoiding direct contact between them.

Blackett, Patrick M. S. Born: July 13, 1897, London, England; died: July 13, 1974. Specialized in nuclear physics and cosmic radiation, and is best known for photographing a nuclear disintegration in 1925, proving the existence of positrons (positive electrons in the nucleus) attained from gamma rays. His experimentation depended on the instrumentation he used, which included the cloud chamber that he had perfected and the Geiger counter. Recipient of the Nobel Prize in Physics for 1948 for his contributions in the exploration of cosmic rays. Together with Giuseppe Occhialini, in 1932, he combined two Geiger counters and a cloud chamber, through which entering cosmic rays were photographed. In fact, charged particles were detected by the Geiger counter while their paths were traced in the cloud chamber.

Bloch, Felix. Born: October 23, 1905, Zurich, Switzerland; died: September 10, 1983, Zurich, Switzerland. Specialized in nuclear physics, and is best known for the analysis of the magnetic properties of a variety of substances through a method based on nuclear magnetic moments induced by nuclear magnetic resonance. Recipient of the Nobel Prize in Physics for 1952, along with Edward Mills Purcell, for his discoveries of the magnetism within the nucleus of the atom, independently of Purcell. Note that such a form of magnetism was discovered in the 1930s. He continued the work of Isidor Rabi (winner of the Nobel of Prize in Physics for 1944), who had devised a method for determining nuclear magnetic moments through induced resonance with electromagnetic waves.

Bloembergen, Nicolaas. Born: March 11, 1920, Dordrecht, Netherlands. Specialized in optics and quantum electronics, and is best known for developing the new field of nonlinear optics. Recipient of the Nobel Prize in Physics for 1981, along with Arthur L. Schawlow and Kai M. Siegbahn, for his work on the response of matter exposed to lasers.

Bohr, Aage. Born: June 19, 1922, Copenhagen, Denmark. Specialized in nuclear physics, and is best known for a new modeling

Chandrasekhar, Subrahmanyan. Nobel Laureate in Physics 1983. Copyright © The Nobel Foundation. Used with permission

Röntgen, Wilhelm Conrad. Nobel Laureate in Physics 1901. Copyright © The Nobel Foundation. Used with permission.

Becquerel, Antoine. Nobel Laureate in Physics 1903. Copyright © The Nobel Foundation. Used with permission.

Curie, Marie. Nobel Laureate in Physices 1903 and Chemistry 1911. Copyright © The Nobel Foundation. Used with permission.

of the composition of the atomic nucleus. Recipient of the Nobel Prize in Physics for 1975, along with Ben R. Mottelson and L. James Rainwater, for his work demonstrating the asymmetry of the nuclear structure due to the vibration and rotation induced by the excitations of the nucleons (protons and neutrons).

Bohr, Niels. Born: October 7, 1885, Copenhagen, Denmark; died: November 18, 1962, Copenhagen, Denmark. Specialized in atomic structure and quantum theory, and is best known for his investigation of atomic structure. Recipient of the Nobel Prize in Physics for 1922 for his work describing the components (electrons and nuclei) of the atom, the interaction between such components, and the emission of radiations through quantum theory (emission of energy in quanta by electrons as they change orbital position).

Born, Max. Born: December 11, 1882, Breslau, Germany; died: January 5, 1970, Gottingen, West Germany. Specialized in quantum mechanics, and is best known for his probability interpretation of the wave function in quantum mechanics. Recipient of the Nobel Prize in Physics for 1954, along with Walther Bothe, for formulating the first comprehensive theory of atomic structure by elaborating Werner Heisenberg's algebraic formulation of quantum mechanics, independently of the wave mechanical formulation of quantum mechanics by Erwin Schrödinger.

Bothe, Walther. Born: January 8, 1891, Oranienburg, Germany; died: February 8, 1957, Heidelberg, West Germany. Specialized in particle physics and nuclear energy, and is best known for his studies of the collisions of photons and electrons. Recipient of the Nobel Prize in physics for 1954, along with Max Born, for discovering that electrons and photons retain the same amount of energy and momentum even after impacting (momentum and energy are conserved).

Bragg, Sir Lawrence. Born: March 31, 1890, Adelaide, South Australia, Australia; died: July 1, 1971, Ipswich, Suffolk, England. Specialized in X-ray crystallography, and is best known for his contribution to the foundation of this discipline, in collaboration with his father, Sir William Henry Bragg. Recipient of the Nobel Prize in Physics for 1915, together with his father, for the mathematical analysis of crystal structures, determining them for zinc blende, diamond, and sodium chloride.

Bragg, Sir William Henry. Born: July 2, 1862, Westward, near Wigton, Cumberland, England; died March 12, 1942, London, England. Specialized in radioactivity, X-ray spectroscopy, and X-ray crystallography, and is best known for his pioneering work in and establishment of X-ray crystallography, in collaboration with his son, Lawrence. Recipient of the Nobel Prize in Physics for 1915, along with son, for the development of the X-ray spectrometer, which measures the strength of an X-ray beam reflected from a crystal face.

Brattain, Walter H. Born: February 10, 1902, Amoy, China; died: October 13, 1987, Seattle, Washington. Specialized in solid-state physics, and is best known for his work on semiconductors. Recipient of the Nobel Prize in Physics for 1956, along with William Shockley and John Bardeen, for his work in collaboration with them that generated the transistor.

Braun, Karl Ferdinand. Born: June 6, 1850, Fulda, Hesse-Kassel, Germany; died: April 20, 1918, Brooklyn, New York. Specialized in wireless telegraphy, and is best known for improving the transmitting and receiving apparatuses originally developed by G. Marconi. Recipient of the Nobel Prize in Physics for 1909, along with Guglielmo Marconi, for reaching longer distances in radio communications by producing, through resonance, higher-intensity radio waves, which he obtained by making modifications to the circuitry of the original transmitter devised by Marconi.

Bridgeman, Percy Williams. Born: April 21, 1882, Cambridge, Massachusetts; died: August 20, 1961, Randolph, New Hampshire. Specialized in high-pressure physics, and is best known for his experimentation on substances subjected to extraordinarily high pressures. Recipient of the Nobel Prize in Physics for 1946 for his investigations of the effects of high pressure on several substances in the solid, liquid, and gaseous states, including ice and heavy water. Achieving pressures that occasionally reached 400,000 atmospheres (5,880,000 psi) in apparatuses of his design, he studied viscosity, heat conduction, electrical resistance, and crystal structures of several materials under such stressful conditions.

Broglie, Louis de. Born: August 15, 1892, Dieppe, France; died: March 19, 1987, Louveciennes, Yvelines, France. Specialized in quantum physics and wave mechanics, and is best known for formulation of the wave theory describing the behavior of atomic particles. Recipient of the Nobel Prize in Physics for 1929 for modeling the thesis that matter (particles) could also behave in a wavelike manner.

Chadwick, Sir James. Born: October 20, 1891, Manchester, England; died: July 24, 1974, Cambridge, England. Specialized in atomic and nuclear physics, and is best known for the determination of the neutron within the atomic nucleus. Recipient of the Nobel Prize in Physics for 1935, for proving the existence of a neutron previously proposed in theory by Ernest Rutherford in 1920 and for formulating a new methodology to determine the mass of the nucleus.

Chamberlain, Owen. Born: July 10, 1920, San Francisco, California. Specialized in nuclear physics, and is best known for the analytical and experimental methodology leading to the discovery of the antiproton, in collaboration with Emilio Segrè. Recipient of the Nobel Prize in Physics for 1959, along with Emilio Segrè, for the joint formulation of the methodology used to discover the antiproton by means of the particle accelerator built by Ernest Orlando Lawrence (winner of the Nobel Prize in Physics, 1939) at the University of California at Berkeley.

Chandrasekhar, Subrahmanyan. Born: October 19, 1910, Lahore, India. Specialized in astrophysics, and is best known for formulating the theory of the white dwarf stars. Recipient of the Nobel Prize in Physics for 1983, along with William Fowler, for his studies on the theory of the evolution of stars, expressing the principle that white dwarf stars have a mass that does not exceed 1.5 times the mass of the sun because stars having larger masses will eventually collapse, becoming neutron stars or so-called black holes.

Cherenkov, Pavel Alekseyevich. Born: July 28, 1904, Novaya Chigla, Russia. Specialized in nuclear physics and particle physics, and is best known for the discovery of the so-called Cherenkov radiation, named after him, which eventually led to the discovery of the antiproton. Recipient of the Nobel Prize in Physics for 1958, along with Ilya Mikhailovich Frank and Igor Yevgenyevich Tamm, for formulating the theory, supported by the other two recipients, that when liquids are bombarded by gamma rays, a glowing phenomenon of light is generated by nuclear particles moving faster than light. Note that this is not contrary to Einstein's theory that the speed of light is the highest possible speed because Einstein referred to the speed of light in a vacuum, whereas its speed in a liquid is much lower than it is in a vacuum.

Cockcroft, Sir John Douglas. Born: May 27, 1897, Todmorden, Yorkshire, England; died: September 18, 1967, Cambridge, England. Specialized in nuclear physics, and is best known for proving the possibility of splitting the atomic nucleus by bombarding it. Recipient of the Nobel Prize in Physics for 1951, along with Ernest Thomas Sinton Walton, for building an accelerator (Cockcroft-Walton accelerator) that at a voltage of 600,000 V produced a beam of protons that generated two nuclei of helium from a thin film of metallic lithium.

Compton, Arthur Holly. Born: September 10, 1892, Wooster, Ohio; died: March 15, 1962, Berkeley, California. Specialized in X-radiation and optics, and is best known for the so-called Compton effect, named after him, proving that when a substance is exposed to X rays it emits two kinds of radiation—one having equal wavelength to the incident X rays and another, secondary, radiation consisting of scattered rays with a different wavelength, larger than the first. Recipient of the Nobel Prize in Physics for 1927, along with Charles Thomson Rees Wilson, for his discovery, which proved for the first time the validity of Einstein's theory of light quanta.

Cooper, Leonn. Born: February 28, 1930, New York, New York. Specialized in superconductivity, and is best known for a comprehensive theory explaining the phenomenon of superconductivity. Recipient of the Nobel Prize in Physics for 1972, along with John

Bardeen and John Robert Schrieffer, for their joint formulation of the theory. In this theory, superconductive materials generate couples of electrons (Cooper pairs) that induce the free electrons to have coordinated motions, whereas in regular materials the electrons maintain random motions.

Cronin, James W. Born: September 29, 1931, Chicago, Illinois. Specialized in particle physics, and is best known for his experimental work on the "neutral K-meson" (subatomic particle). He found that the decay of such a particle could happen in an asymmetrical manner, which proved that symmetry is not an absolute requirement in physics. Recipient of the Nobel Prize in Physics for 1980, along with Val L. Fitch, for their joint work that discovered that two K-mesons out of a thousand decayed without symmetry, implying the absence of symmetry with matter and antimatter.

Curie, Marie. Born: November 7, 1867, Warsaw, Poland; died: July 4, 1934, Sancellemoz, near Sallanches, France. Specialized in radioactivity, and is best known for the discovery of polonium and radium, together with her husband, Pierre Curie, on the basis of Antoine-Henri Becquerel's findings about the spontaneous radioactivity in uranium. Recipient of the Nobel Prize in Physics for 1903, along with Pierre Curie and Antoine-Henri Becquerel, for the discovery of the existence of these two new radioactive elements.

Curie, Pierre. Born: May 15, 1859, Paris, France; died: April 19, 1906 in Paris. Specialized in radioactivity, magnetism, and crystallography, and is best known for the discovery of polonium and radium, together with his wife, Marie Curie, on the basis of Antoine-Henri Becquerel's findings about the spontaneous radioactivity in uranium. Recipient of the Nobel Prize in Physics for 1903, along with Marie Curie and Antoine-Henri Becquerel, for the discovery of the existence of these two new radioactive elements.

Dalén, Nils Gustaf. Born: November 30, 1869, Stenstorp, Sweden; died: December 9, 1937, Lidingo, Sweden. Specialized in engineering, and is best known for inventing a lighting system based on the burning of acetylene gas, used for lighthouses and buoys worldwide. Recipient of the Nobel Prize in Physics for 1912 for a gaslight system using an explosion-proof porous mass containing the required acetylene. Installed on buoys and lighthouses, the lights required refueling approximately once a year, conserving their fuel by flashing rather than being constantly lit, and by automatically turning off during daylight and turning on at night.

Davisson, Clinton Joseph. Born: October 22, 1881, Bloomington, Illinois; died: February 1, 1958, Charlottesville, Virginia. Specialized in electron physics, and is best known for discovering the diffraction of electrons, similar to the diffraction of X rays with wave properties, in accordance with the theory previously established by Louis de Broglie. Recipient of the Nobel Prize in Physics for 1937, along with George P. Thomson, for his studies on a beam of low-energy electrons being scattered from the surface of a nickel crystal, following the patterns of X rays.

Dehmelt, Hans Georg. Born: September 9, 1922, Görlitz, Germany. Presently at U. of Washington, Seattle. Recipient of one-half of the Nobel Prize in Physics for 1989, together with Wolfgang Paul, for their joint work on the development of the ion trap technique. The other half of the prize was assigned to Norman Ramsey for the invention of the separated oscillatory fields method and its use in the hydrogen maser and other atomic clocks.

Dirac, Paul Adrien Maurice. Born: August 8, 1902, Bristol, Gouchestershire, England; died: October 20, 1984, Tallahassee, Florida. Specialized in quantum mechanics, and is best known for the formulation of a relativistic wave equation describing the properties of the electron's spin. Recipient of the Nobel Prize in Physics for 1933, along with Erwin Schrödinger, for hypothesizing the existence of the positron, which was confirmed experimentally.

Einstein, Albert. Born: March 14, 1879, Ulm, Württemberg, Germany; died: April 18, 1955, Princeton, New Jersey. Specialized in theoretical physics, and is best known for formulation of the theory of relativity (special and general) and for his work on Brownian motion. Recipient of the Nobel Prize in Physics for 1921 for the light quantum and photoelectric effect (not for his theories of relativity).

Esaki, Leo. Born: March 12, 1925, Osaka, Japan. Specialized in quantum mechanics and solid-state physics, and is best known for his invention of the Esaki tunnel diode. Recipient of the Nobel Prize in Physics for 1973, along with Ivar Giaever and Brian D. Josephson, for opening the field of tunneling research with his experimentation.

Fermi, Enrico. Born: September 29, 1901, Rome, Italy; died: November 29, 1954, Chicago, Illinois. Specialized in radioactivity and nuclear reactions, and is best known for the bombardment of the nucleus of atoms with neutrons, consequently changing one element into another new element, and for his contribution to the development of the atomic bomb. He succeeded in attaining what alchemists had aimed to achieve during the Middle Ages, when they tried to convert metals into another metal (gold). He proved that new elements could be artificially made, in addition to the 92 elements of the periodic table. By bombarding uranium he attained two additional elements, "Ausenium" and "Hesperium," respectively elements 93 and 94 in the periodic table. Recipient of the Nobel Prize in Physics for 1938 for discovering the statistical laws of atomic particles and electrodynamic spectroscopy, which led to the discovery of the possibility of nuclear bombardment with neutrons.

Feynman, Richard P. Born: May 11, 1918, New York, New York; died: February 15, 1988, Los Angeles, California. Specialized in quantum electrodynamics, and is best known for his theory of quantum electrodynamics, and the interrelationships of subatomic particles (electrons, positrons, and photons). Recipient of the Nobel Prize in Physics for 1965, along with Shin'ichirō Tomonaga and Julian Schwinger, for reconstructing quantum mechanics and electrodynamics through a graphical representation, referred to as the Feynman diagrams.

Fitch, Val L. Born: March 10, 1923, Merriman, Nebraska. Specialized in particle physics, and is best known for his experimental work on the "neutral K-meson" (subatomic particle). It was found that the decay of such a particle could happen in an asymmetrical manner, which proved that symmetry is not an absolute requirement in physics. Recipient of the Nobel Prize in Physics for 1980, along with James W. Cronin, for their joint work, which discovered that two K-mesons out of a thousand decayed without symmetry, implying the absence of symmetry with matter and antimatter.

Fowler, William A. Born: August 9, 1911, Pittsburgh, Pennsylvania. Specialized in astrophysics and nuclear physics, and is best known for his work in nuclear reactions that take place in stars. Recipient of the Nobel Prize in Physics for 1983, along with Subrahmanyan Chandrasekhar, for demonstration of the formation of energy in the interior of stars and the formation of elements in the universe.

Franck, James. Born: August 26, 1882, Hamburg, Germany; died: May 21, 1964, Göttingen, West Germany. Specialized in atomic and molecular physics, and together with Gustav Hertz, is best known for experimentation on the impact of the collision of an electron with an atom. Recipient of the Nobel Prize in Physics for 1925, along with Gustav Hertz, for their joint experiments that established the theory of collisions between atoms and electrons, which led the way to further understanding of the structure of atoms and molecules, as well as for verifying the quantum theory for the energy of the atom.

Frank, Ilya Mikhailovich. Born: October 23, 1908, St. Petersburg, Russia. Specialized in nuclear physics, particle physics, and optics, and best known for his theory explaining Cherenkov radiation. Recipient of the Nobel Prize in Physics for 1958, along with Igor Yevgenyvich Tamm and Pavel Alekseyevich Cherenkov, for his work, together with Tamm, on the explanation and mathematical model of the phenomenon credited to Cherenkov.

Gabor, Dennis. Born: June 5, 1900, Budapest, Hungary; died: February 8, 1979, London, England. Specialized in electron optics and holography, and is best known for this discovery of the principles of holography (three-dimensional imaging). Recipient of the Nobel Prize in Physics for 1971 for his work on holography completed in 1940. Note that only after the invention of the laser, did holography find some practical applications. Holography is the three-dimensional representation of images based on two fundamental elements; one is the ability to record an image in terms of

Michelson, Albert Abraham. Nobel Laureate in Physics 1907. Copyright © The Nobel Foundation. Used with permission.

Marconi, Guglielmo. Nobel Laureate in Physics 1909. Copyright © The Nobel Foundation. Used with permission.

lanck, Max K. E. L. Nobel Laureate in Physics 1918. Copyright © The Nobel Foundation. 'sed with permission.

Einstein, Albert. Nobel Laureate in Physics 1921. Copyright © The Nobel Foundation. Used with permission.

the differences in intensity of the light reflected by the individual points of an object, and the other is the characteristics of the phase of the light being reflected from different points.

Friedman, Jerome Born: March 28, 1930, Chicago, Illinois, USA. Specialized in nuclear physics, and is best known for his work in collaboration with the Stanford Linear Accelerator Center (SLAC), while working at MIT. The experimentation referred to as "SLAC-MIT" involved the use of the two-mile-long linear accelerator at Stanford, in which the structure of nucleons (protons and neutrons) was studied using record-high-energy electrons as probes. Recipient of the Nobel Prize in Physics for 1990, along with Henry Kendall and Richard Taylor, for their joint work in pioneering investigations concerning deep inelastic scattering of electrons on protons and bound neutrons, which have been of essential importance for the development of the quark model in particle physics.

Gell-Mann, Murray. Born: September 15, 1929, New York, New York. Specialized in particle physics, and is best known for his theories and classifications of subatomic particles, applicable to those already known as well as those discovered later, such as the pi-mesons and the omega-minus. Recipient of the Nobel Prize in Physics for 1989 for his theory of "strangeness" and for the theory of quarks. The term "quark" applies to particles that are assumed to be the basic components of all other subatomic particles.

Giaever, Ivar. Born: April 5, 1929, Bergen, Norway. Specialized in quantum mechanics, solid-state physics, and biophysics, and is best known for his work on electron tunneling in superconductors. Recipient of the Nobel Prize in Physics for 1973, along with Leo Esaki and Brian D. Josephson, for his work on electron tunneling in superconductors following Esaki's work on electron tunneling in semiconductors.

Glaser, Donald A. Born: September 21, 1926, Cleveland, Ohio. Specialized in particle physics, and is best known for his invention of the bubble chamber, based on analysis of the passage of a high-energy atomic particle that will produce the formation of bubbles in a liquid heated to just below the boiling point. Recipient of the Nobel Prize in Physics for 1960 for the invention of the bubble chamber, which followed the cloud chamber invented by Charles Thomson Rees Wilson, used for tracing radioactive decay products in low-energy motions.

Glashow, Sheldon L. Born: December 5, 1932, New York, New York. Specialized in particle physics, and is best known for theory relating the electromagnetic force and the weak force of the atomic nucleus. Recipient of the Nobel Prize in Physics, along with Steven Weinberg and Abdus Salam, for that theory and for discovery of the phenomenon of the "weak neutral current" developed when an electron changes into a neutrino, which is in turn changed back to an electron.

Guillaume, Charles-Édouard. Born: February 15, 1861, Fleurier, Switzerland; died June 13, 1938, Sevres, France. Specialized in metallurgy and metrology, and is best known for the discovery of "invar" and "elinvar," special alloys for high-precision instrumentations. Recipient of the Nobel Prize in Physics for 1920 for his research in metallurgy and his discovery of "invar," with an extremely low coefficient of thermal expansion and for the discovery of "elinvar," with extremely low changes in the coefficient of elasticity with respect to temperature variation. The use of such alloys enabled many precision instruments, such as chronometers and geodesic apparatuses, to be built with acceptable tolerance in their measurements.

Heisenberg, Werner. Born: December 5, 1901, Wurzburg, Germany; died: February 1, 1976, Munich, West Germany. Specialized in quantum mechanics, and is best known for his theory of the uncertainty principle (Heisenberg theory), which contributed to the development of quantum mechanics. Recipient of the Nobel Prize in Physics for 1932 for his discoveries in quantum mechanics that established the impossibility of determining the position and velocity of a particle because, with efforts to establish the position of a particle, its velocity becomes more uncertain, and vice versa.

Hertz, Gustav. Born: July 22, 1887, Hamburg, Germany; died: October 30, 1975, Berlin, East Germany. Specialized in atomic and molecular physics, and is best known for his experimental work in conjunction with James Franck on light emissions from ionized mercury vapor, proving the proposed structure of the Bohr model of the atom and the level of its energy states. Recipient of the Nobel Prize in Physics for 1925, along with James Franck, for their work on methodologies for studying the elastic collisions of electrons and ions, atoms and molecules.

Hess, Victor Franz. Born: June 24, 1883, Waldstein Castle, Near Graz, Styria, Austria; died: December 17, 1964, Mount Vernon, New York. Specialized in cosmic radiation, and is best known for the discovery of cosmic radiation. Recipient of the Nobel Prize for Physics for 1936, along with Carl David Anderson, for his discovery of this new radiation never before suspected. Searching for the source of radioactivity, Hess discovered the existence of past radiations (cosmic rays) that must come from outer space, probably past the known galaxies, because they definitely are not generated by the sun or by any specific stars. Experimentally, Hess established that the intensity of cosmic rays increases with height, being doubled at three miles above the earth and eventually increasing much more, as further experimentation later demonstrated.

Hewish, Antony. Born: May 11, 1924, Fowey, Cornwall, England. Specialized in radio astronomy, and is best known for the discovery of the so-called pulsars, the final stage in the evolution of certain stars. Recipient of the Nobel Prize in Physics for 1974, along with Martin Ryle, for the discovery of pulsars, which he detected in their joint work, using a radio telescope of their design. Extremely small, with a diameter of only 10 kilometers, consisting of densely concentrated neutrons and surrounded by a strong magnetic field, pulsars emit radio pulses that had been captured by their telescope. Pulsars eventually were found to be a final state in the evolution of stars.

Hofstadter, Robert. Born: February 5, 1915, New York, New York. Specialized in nuclear physics, and is best known for his pioneering work in analyzing the structure of atoms attained by bombarding the atomic nucleus with highly energized electrons. Recipient of the Nobel Prize in Physics for 1961, along with Rudolf Mossbauer, for his determination of the structure of the nucleus and the distribution of charges within it, through experimental methods that he devised.

Jensen, J. Hans D. Born: June 25, 1907, Hamburg, Germany; died: February 11, 1973, Heidelberg, West Germany. Specialized in nuclear physics, and is best known for the development of the shell model that illustrated the structure of the nucleus in an innovative manner. Recipient of the Nobel Prize in Physics for 1963, along with Eugene Wigner and Maria Goeppert Mayer, for his discoveries of the structure of the nucleus, which coincided with similar findings by cowinner Maria Goeppert Mayer. This new model of the nucleus, which replaced the "liquid drop" model, explained the motions of protons and neutrons in the nucleus and proved the existence of the so-called magic numbers (2, 8, 20, 28, 50, 82, 126), representing the numbers of neutrons or protons present in highly stable elements and the isotopes of such elements.

Josephson, Brian D. Born: January 4, 1940, Cardiff, Glamorgan, Wales. Specialized in quantum mechanics and solid-state physics, and is best known for the theory of tunneling through two superconductors. Recipient of the Nobel Prize in Physics for 1973, along with Leo Esaki and Ivar Giaever, for his contribution, which complemented the work of the two cowinners of the prize. Specifically, his theory explained the tunneling phenomenon, in which the barrier between two superconductors is penetrated by a supercurrent even in the absence of voltage. However, if a constant voltage is applied, an alternating current with a high frequency will pass through the barrier (Josephson Effect).

Kamerlingh Onnes, Heike Born: September 21, 1853, Groningen, Netherlands; died February 21, 1926, Leiden, Netherlands. Specialized in low-temperature physics, and is best known for his experimentation on the liquefaction of gases—specifically, his success with the liquefaction of helium. Recipient of the Nobel Prize in Physics in 1913 for his success in liquefying helium in 1908. Although most other gases already had been liquefied in a laboratory, helium, newly discovered in 1895, had not yet been successfully liquefied. His experimentation succeeded with helium by attaining particularly low temperatures and sustaining them for a considerable time, allowing the determination of several physical characteristics, including the reduction of electric resistance at such low temperatures.

Kapitsa, Pyotr Leonidovich. Born: July 9, 1894, Kronshtadt, Russia; died April 8, 1984, Moscow, Soviet Union. Specialized in low-temperature physics and plasma physics, and is best known for the liquefaction of gases (helium and air). Recipient of the Nobel Prize in Physics for 1978, along with Arno A. Penzias and Robert W. Wilson, for his contribution to the liquefaction in gases and discovery of the superfluidity of helium. Having invented an apparatus for the liquefaction of helium at a temperature of 2.2 degrees kelvin, he had discovered a means of mass production of liquid helium and liquid air. As part of his discovery, he proved that at such a temperature, liquid helium loses all its viscosity.

Kastler, Alfred. Born: May 3, 1902, Guebwiller, Alsace, Germany; died: January 7, 1984, Bandol, France. Specialized in optical spectroscopy and Hertzian resonances, and best known for his discoveries in 1950 and 1952 that led to the invention of masers and lasers by Townes in the United States and Prokhorov and Basov in the Soviet Union. Recipient of the Nobel Prize in Physics for 1966 for his optical method applied to the study of Hertzian resonances and for the development of an optical pumping apparatus for detecting them.

Kendall, Henry Born: December 9, 1926, Houston, Texas, USA. Specialized in nuclear physics, and is best known for his work in collaboration with the Stanford Linear Accelerator Center (SLAC), while working at MIT. The experimentation referred to as "SLAC-MIT" involved the use of the two mile-long linear accelerator at Stanford, in which the structure of nucleons (protons and neutrons) was studied using record-high-energy electrons as probes. Recipient of the Nobel Prize in Physics for 1990, along with Jerome Friedman and Richard Taylor, for their joint work in pioneering investigations concerning deep inelastic scattering of electrons on protons and bound neutrons, which have been of essential importance for the development of the quark model in particle physics.

Klitzing, Klaus Von. Born: June 28, 1943, Schroda, Germany. Specialized in condensed-matter physics, and is best known for the quantum Hall effect, and its application to semiconductors, mainly used in computer technology. Recipient of the Nobel Prize in Physics for 1985 for his work on the Hall effect on semiconductors near absolute zero temperature, observing that variation of the magnetic field as a function of voltage and current did not occur smoothly but varied abruptly under the influence of the charge of the electron and Planck's constant.

Kusch, Polykarp. Born: January 26, 1911, Blankenburg, Germany. Specialized in atomic and molecular physics, and is best known for the high precision of his measurements of the magnetic strength of the electron. Recipient of the Nobel Prize in Physics for 1955, along with Willis Lamb, for his work at Columbia University, independent from Willis, who was also working at the same institution and reached the same conclusion. In checking Dirac's theory and finding a deviation of one per thousand, he was able to establish the limitations of that theory and paved the way for the formulation of quantum electrodynamics.

Lamb, Willis Eugene, Jr. Born: July 12, 1913, Los Angeles, California. Specialized in quantum electrodynamics, and is best known for his work on the atomic structure of hydrogen, both theoretical and experimental, which led to a restructuring of the theory of quantum electrodynamics. Recipient of the Nobel Prize in Physics for 1955, along with Polykarp Kusch, for reinterpreting the explanation of the "fine" structure of the hydrogen atom. Note that this "fine" structure consists of the energy levels of the various orbits of the electron in the atom, so grouped that neighboring energy levels are widely spaced.

Landau, Lev Davidovich. Born: January 22, 1908, Baku, Azerbaijan, Russian Empire; died: April 1, 1968, Moscow, Soviet Union. Specialized in quantum mechanics, and is best known for his understanding of liquid helium in the superfluid state, which enhanced scientists' knowledge of the properties of quantum liquids. Recipient of the Nobel Prize in Physics for 1962 for his applications of quantum mechanics methods, which led him to a theory of large-scale quantum behavior.

Laue, Max von. Born: October 9, 1879, Pfaffendorf, near Koblenz, Germany; died: April 23, 1960, Berlin, Germany. Is best known for discovering the diffraction of X rays in penetrating crystals. Recipient of the Nobel Prize in Physics for 1914 for discovering the phenomenon of X-ray diffraction, which eventually created X-ray crystallography and X-ray spectroscopy. Note that his discovery enabled several important deductions to be made. X rays were found to have the same electromagnetic properties as light except for their wavelength, which is 10,000 times shorter than that of light. Also, through such a discovery it was possible to locate the position of atoms within the structure of crystals.

Lawrence, Ernest Orlando. Born: August 8, 1901, Canton, South Dakota; died: August 27, 1958, Palo Alto, California. Specialized in nuclear physics, and is best known for the invention of the cyclotron and for his contribution to the development of the atomic bomb. Recipient of the Nobel Prize in Physics for 1939 for the invention of the cyclotron, a device used to accelerate ions to high-energy levels for the bombardment of atoms. Its evolution proceeded through various steps, starting from a 12-inch-diameter machine, and reaching the dimensions of a 184-inch-diameter machine, called the synchrotron, capable of accelerating ions to energies of several billion electron volts.

Lederman, Leon M. Born: July 15, 1922, New York, New York. Specialized in high-energy particle physics, and is best known for the neutrino-beam artificially produced in the laboratory, along with the cowinners of the prize. Recipient of the Nobel Prize in Physics for 1988, along with Jack Steinberger and Melvin Schwartz, for their joint work on the so-called weak interaction, defined as one of the primary forces of nature (gravitational, electromagnetic, strong, and weak). Using neutrino beams produced for the first time in the laboratory, they discovered a new type of neutrino, advancing the theory of the "standard model" in the field of particle physics.

Lee, Tsung-Dao. Born: November 25, 1926, Shanghai, China. Specialized in particle physics and statistical mechanics, and is best known for his suggestions, together with the cowinner of the prize, concerning theory and experimentation proving the absence of conservation of parity. Recipient of the Nobel Prize in Physics for 1957, along with Chen Ning Yang, for their joint work on the theoretical aspects of the law of conservation of parity, which led to later discoveries proving that parity was not conserved.

Lenard, Philipp. Born: June 7, 1862, Pozsony (Pressburg), Hungary; died: May 20, 1947, Messelhausen, Germany. Specialized in photoelectricity and electrons, and is best known for his studies of cathode rays. Recipient of the Nobel Prize in Physics for 1905 for devising a method to induce cathode rays to pass through a so-called window from a tube of rarefied gas to the open air, allowing an in-depth study of the cathode rays.

Lippman, Gabriel. Born: August 16, 1845, Hollerich, Luxembourg; died July 13, 1921, at sea, en route from Canada to France. Specialized in applied mathematical physics, and is best known for theoretical and applied work that generated color photography. Recipient of the Nobel Prize in Physics for 1908 for presenting a photographic process that recorded a colored photograph on a single plate in a single exposure, thus advancing the state of the art in color photography.

Lorentz, Hendrik Antoon. Born: July 18, 1853, Arnhem, Netherlands; died: February 4, 1928, Haarlem, Netherlands. Specialized in electromagnetic theory, and is best known for his theory explaining the Zeeman effect. Recipient of the Nobel Prize in Physics for 1902, along with Pieter Zeeman, for his work in cooperation of the cowinner of the prize, whose experimental discovery—named after him—consisted of the splitting of the lines of the spectrum when the source was exposed to a magnetic field.

Marconi, Guglielmo. Born: April 25, 1874, Bologna, Italy; died: July 20, 1937, Rome, Italy. Specialized in radiotelegraphy, and is best known for wireless telegraphy, which made transatlantic communication possible. Recipient of the Nobel Prize in Physics for 1909, along with Karl Braun, for his initial and subsequent transmission through Hertzian waves, which gradually increased in distance. Notice that Marconi's results were eventually improved on by the cowinner of the prize, who succeeded in obtaining stronger signals, overcoming the "damped oscillation" phenomenon.

Mayer, Maria Goeppert. Born: June 28, 1906, Kattowitz, Upper Silesia, German; died: February 20, 1972, San Diego, California. Specialized in nuclear physics, and is best known for a shell model of the atomic nucleus, formulated independently of J. Hans D. Jensen. Recipient of the Nobel Prize in Physics for 1963, along with J. Hans D. Jensen, for her work explaining the effect of the "magic numbers" on several properties of the atomic nucleus. Such numbers (2, 8, 20, 28, 50, 82, 126), indicating the number of protons or neutrons in a nucleus, characterize a condition of strong stability when they occur.

Michelson, Albert Abraham. Born: December 19, 1852, Strelno, Prussia; died May 9, 1931, Pasadena, California. Specialized in optics, spectroscopy, and interferometry, and is best known for the "Michelson–Morley experiment" on measuring the effects of the earth's orbital motion on the speed of light. Recipient of the Nobel Prize in Physics for 1907 for the invention of optical precision instruments and for their use in metrology and spectroscopy. Among them is the interferometer, which allowed measurements up to 100 times more accurate than those that had been possible before with the most accurate microscope.

Millikan, Robert Andrews. Born: March 22, 1868, Morrison, Illinois; died: December 19, 1953, San Marino, California. Specialized in the electronic charge and the photoelectric effect, and is best known for his experiments on the unit charge of the electron. Recipient of the Nobel Prize in Physics for 1923 for his experimental work on the charge of the electron and the photoelectric effect. With this work, he confirmed the validity of Einstein's equation for the photoelectric effect.

Mossbauer, Rudolf Ludwig. Born: January 31, 1929, Munich, Germany. Specialized in gamma radiation, and is best known for his work on nuclear gamma radiations and the discovery of the Mossbauer effect. Recipient of the Nobel Prize in Physics for 1961, along with Robert Hofstadter, for his work on the resonance emission and absorption of nuclear gamma radiation. Part of his contributions included the formulation of a theory and the devising of an experimental system for studying such resonances.

Mott, Sir Nevill. Born: September 30, 1905, Leeds, Great Britain. Specialized in solid-state physics, and is best known for his work on the differentiation of conductors, semiconductors, and insulators. Recipient of the Nobel Prize in Physics for 1977, along with Philip W. Anderson and John H. Van Vleck, for his work leading to the "Mott transitions"and to the "Mott–Anderson transition" theories.

Mottelson, Ben R. Born: July 9, 1926, Chicago, Illinois. Specialized in nuclear physics, and is best known for the formulation of a comprehensive theory of nuclear behavior. Recipient of the Nobel Prize in Physics for 1975, along with L. James Rainwater and Aage Bohr, for his work in collaboration with the cowinners of the prize. Specifically, he, together with Aage Bohr, experimentally proved the theory formulated by Rainwater. This theory envisioned a configuration different from a sphere but deformed in a more oblong shape under the action of centrifugal force determined by the motion of the nucleus itself. The theory departed from the two theories previously proposed: the liquid drop theory and the shell model theory.

Muller, Karl Alexander. Born: April 20, 1927, Basel, Switzerland. Specialized in solid-state physics and superconductivity, and is best known for discovering superconductivity in a ceramic material at a temperature much higher than any previously discovered. Recipient of the Nobel Prize in Physics for 1987, along with J. Georg Bednorz, for his work in collaboration with the cowinner on superconductivity. This discovery opened this field to international research on other, more efficient materials for use as superconductors.

Néel, Louis-Eugène-Felix. Born: November 22, 1904, Lyons, France. Specialized in nuclear magnetism, and is best known for experimental work in the field of magnetism. Recipient of the Nobel Prize in Physics for 1970 along with Hannes Alfvén, for his discovery of ferromagnetic and antiferromagnetic materials, which led to the clarification of several principles in magnetism. As a consequence of findings, significant progress was made in various technological fields, including communications equipment, computer data storage, and so on.

Bohr, Niels. Nobel Laureate in Physics 1922. Copyright © The Nobel Foundation. Used with permission.

Hertz, Gustav. Nobel Laureate in Physics 1925. Copyright © The Nobel Foundation. Used with permission.

De Broglie, Prince Louis-Victor. Nobel Laureate in Physics 1929. Copyright © The Nobel Foundation. Used with permission.

Schrödinger, Erwin. Nobel Laureate in Physics 1933. Copyright © The Nobel Foundation. Used with permission.

Paul Wolfgang Born: , 1913, Federal Republic of Germany. Presently at the U. of Bonn. Recipient of one-half of the Nobel Prize in Physics for 1989, together with Hans Dehmelt, for their joint work on the development of the ion trap technique. The other half of the prize was assigned to Norman Ramsey for the invention of the separated oscillatory fields method and its use in the hydrogen maser and other atomic clocks.

Pauli, Wolfgang. Born: April 25, 1900, Vienna, Austria; died: December 15, 1958, Zurich, Switzerland. Specialized in quantum mechanics, and is best known for his theory on quantum numbers (the exclusion principle) and for his theories on electrical conductivity in metals and the magnetic properties of matter. Recipient of the Nobel Prize in Physics for 1945 for determination of the requirements for specifying the properties of the orbits of electrons, including energy (such properties had to be identified by four quantum numbers); and for the principle of exclusion, stating that each electron has a different set of quantum numbers.

Penzias, Arno A. Born: April 26, 1933, Munich, Germany. Specialized in radio astronomy, and is best known for his observation of the cosmic microwave background radiation that accompanied the so-called big bang at the creation of the universe. Recipient of the Nobel Prize in Physics for 1978, along with Robert W. Wilson and Pytor L. Kapitsa, for his work conducted jointly with Wilson. Those two had been the first investigators to observe the residue of the radiations that accompanied the explosion at the beginning of the universe. In their experimentation on radio radiation in the galaxy, they detected an unknown radiation of equal intensity in all directions. Such radiations were confirmed by others to be remnants of the original radiations that had occurred 15 billiion years earlier, at the time of the explosion at the birth of the universe, now cooled down enough to be detectable as radio waves.

Perrin, Jean-Baptiste. Born: September 30, 1870. Lille, France; died April 17, 1942, New York, New York. Specialized in molecular physics, and is best known for determining Avogadro's number through various experiments. Recipient of the Nobel Prize in Physics for 1926 for his work on the Brownian motion of particles in an emulsion and for his methodologies for determining Avogadro's number for various substances. Note: Avogadro's number indicates the number of molecules present in a certain amount of a substance.

Planck, Max. Born: April 23, 1858, Kiel, Germany; died: October 3, 1947, Göttingen, West Germany. Specialized in quantum physics, and is best known for the Planck constant and the Avogadro constant. Recipient of the Nobel Prize in Physics for 1918 for his mathematical work on quantum theory. Note that the emission of energy by atoms occurs in "bundles" called quanta. Also note that Planck was the first scientist to create a formula for a general radiation law, which in fact contains the famous (Planck) constant named after him.

Powell, Cecil Frank. Born: December 5, 1903, Tonbridge, Kent, Great Britain; died: August 9, 1969, near Milan, Italy. Specialized in nuclear physics and cosmic radiation, and is best known for the discovery of new elementary particles in cosmic radiation. Recipient of the Nobel Prize in Physics for 1950 for his particle trace analysis in cosmic radiation, using photographic emulsion methods to record the tracks of particles, and for an apparatus capable of furnishing precise quantitative data from the recorded tracks.

Prokhorov, Aleksandr Mikhailovich. Born: July 11, 1916, Atherton, Queensland, Australia. Specialized in quantum radiophysics and quantum electronics, and is best known for his invention of quantum microwave amplification devices (masers) and light amplifiers (lasers) that depend on stimulated emission of radiation, predicted by Einstein in theoretical terms. Recipient of the Nobel Prize in Physics for 1964, along with Nikolay Gennadiyevich Basov and Charles Townes, for producing the first maser in the Soviet Union. A typical example, is demonstrated by the ruby laser, which is a powerful beam of monochromatic, coherent light (when the crests of the light waves coincide) that emerges from a ruby crystal. To produce it, a xenon flash lamp is used to create an inverted population in the ruby, in which a majority of the chromium atoms are set in a high-energy state and a minority of them are put in a ground state (zero energy). At the ground state these atoms generate photons, which stimulate radiations. Such radiations, in turn, are reflected by two face-to-face mirrors that cause the radiations to cross the ruby numerous times. Under these conditions, a ruby laser eventually is obtained.

Purcell, Edward Mills. Born: August 30, 1912, Taylorville, Illinois. Specialized in nuclear magnetic resonance, and is best known for his determination of the magnetic moment of the nucleus, which is different for each element's nucleus. Recipient of the Nobel Prize in Physics for 1952, along with Felix Bloch, for the design and construction of microwave equipment that, through resonance, determined the magnetic moment of a nucleus. A practical application of this work allows the identification of chemical substances because each element's nucleus is characterized by an individual magnetic moment.

Rabi, Isidor Isaac. Born: July 29, 1898, Rymanow, Austria-Hungary; died: January 11, 1988, New York, New York. Specialized in nuclear physics, and best known for measuring the magnetic moment of atoms through a method (resonance method), devised by him, that is now the most widely used technique in modern research. Recipient of the Nobel Prize in physics for 1944 for his experimental work on the behavior of atoms exposed to a magnetic field. Note that experiments on molecular and atomic beams in current use are indeed based on the fundamental method that Rabi devised.

Rainwater, L. James. Born: December 9, 1917, Council, Idaho; died: May 31, 1986, Yonkers, New York. Specialized in structural nuclear physics, and is best known for his work on the determination of the physical shape of the atomic nucleus. Recipient of the Nobel Prize in Physics for 1975, along with Ben R. Mottelson and Aage Bohr, for his theory, subsequently confirmed through experimental work, of the deformed shape of the atomic nucleus during its accelerated motion in a cyclotron.

Raman, Sir Chandrasekhara Venkata. Born: November 7, 1888, Trichinopoly (Tiruchirapalli), India; died: November 21, 1970, Bangalore, India. Specialized in optics, and is best known for the Raman effect, observed experimentally when a monochromatic light beam passes through a transparent substance and is scattered.

Recipient of the Nobel Prize in Physics for 1930 for the "Raman effect," named after him. Note that the Raman effect is attributed to the loss or the gain of photon energy as a result of the light's interaction with the molecules of the medium through which it passes.

Ramsey, Norman Foster. Born: August 27, 1915, Washington, D.C. Presently at Harvard University. Recipient of one-half of the Nobel Prize in Physics for 1989, for the invention of the separated oscillatory fields method and its use in the hydrogen maser and other atomic clocks. The other half of the prize was assigned to Hans Denmelt and Wolfgang Paul for their joint work on the development of the ion trap technique.

Rayleigh, Lord. Born: November 12, 1842, Langford Grove, near Maldon, Essex, England; died: June 30, 1919, Terling Place, Witham, Essex, England. Specialized in acoustics and optics, and is best known for the discovery of argon, one of the noble gases. Recipient of the Nobel Prize in Physics for 1904 for the discovery of argon gas present in the atmosphere. Note that the discovery of the existence of this noble gas was made while Rayleigh was investigating the difference between synthetic nitrogen and nitrogen under natural conditions, as it exists in the atmosphere.

Richardson, Sir Owen Willans. Born: April 26, 1879, Dewsbury, Yorkshire, England; died: February 15, 1959, Alton, Hampshire, England. Specialized in thermionics, and is best known for his work in thermionics (a name formulated by Richardson, himself, to designate the emission of electrons from the hot filaments of metals). Recipient of the Nobel Prize in Physics for 1928 for his theory correlating the emission of electrons and the temperature of the metal providing the emission.

Richter, Burton. Born: March 22, 1931, Brooklyn, New York. Specialized in particle physics, and is best known for the discovery of the J-psi subatomic particle, which weighs three times more than a proton. Recipient of the Nobel Prize in Physics in 1976, along with Samuel C. C. Ting, for his discovery, independent of Ting, of the J-psi particle, at Stanford University. Working in collaboration with other researchers at Stanford University and the University of

California at Berkeley, he had employed a technique based on the collision of electrons and positrons using the Stanford Linear Accelerator. Note that the name for the new particle, simultaneously discovered by Richter and Ting, is a result of combining the name "psi," given to the particle by Richter, and the name "J," given to it by Ting.

Rohrer, Heinrich. Born: June 6, 1933, Buchs, St. Gallen, Switzerland. Specialized in scanning tunneling microscopy, and is best known for his studies on the development of high-technology instruments capable of probing the structure of matter on the scale of atomic and subatomic particles. Such instrumentation implied new theoretical concepts that transcended the conventional visual exploration of form using light; for when the dimensions of the structures to be explored are smaller than the wavelength of light, optical instrumentation is no longer possible and the system used to explore forms must be based on another medium. The ultramicroscope, which is still based on light as the medium of exploration, cannot be used for atomic and subatomic observations. Recipient of the Nobel Prize in Physics for 1986, along with Gerd Binnig and Ernst Ruska, for the development of scanning tunneling microscopy. He and Binnig had worked together on the development of proper instrumentation in the same laboratory in Zurich; their work capitalized on Ernst Ruska's earlier discovery of the electron microscope, developed in 1930, adding substantial refinements 56 years later. In this process, a very fine needle with a sensitive tip—so fine as to reach atomic dimensions—can explore surfaces with an accuracy on the atomic scale. To maintain such accuracy of measurement, the instrumentation includes the interposition of an electric layer between the needle and the surface, avoiding direct contact between them.

Röntgen, William Conrad. Born: March 27, 1845, Lennep, Prussia; died: February 10, 1923, Munich, Germany. Specialized in X-radiation, and is best known for his work on X rays, also called Röntgen rays after him. Recipient of the Nobel Prize in Physics for 1901 for the discovery of X rays and their penetration through substances not permeable to light—a discovery not totally understood at that time, but eventually explored in greater detail by scientists who followed him.

Rubbia, Carlo. Born: March 31, 1934, Gorizia, Italy. Specialized in high-frequency particle physics, and is best known for producing the W and Z particles. Recipient of the Nobel Prize in Physics for 1984, along with Simon van der Meer, for the discovery of the W and Z particles in collaboration with the cowinner. Such particles had been predicted much earlier by three Nobel Prize winners: Glashow, Salam, and Wienberg. The new discovery by Rubbia and van der Meer resulted from teamwork conducted at a research center (Centre Européen de Recherche Nucleaire) sponsored by thirteen European nations. Beams of protons and antiprotons generated in the Super Proton Synchrotron collide with each other, and W and Z particles are produced from the collision. Protons and antiprotons, being particles with opposite charges, travel in circular paths in opposite directions and eventually collide as their paths intersect.

Ruska, Ernst. Born: December 25, 1906, Heidelberg, Germany; died: May 30, 1988, Berlin, West Germany. Specialized in electrical engineering and electron microscopy, and is best known for his work leading to the discovery of the electron microscope. Recipient of the Nobel Prize in Physics for 1986, along with Heinrich Rohrer and Gerd Binnig, for his discoveries in the 1930s of the original electron microscope, in which electron beams were substituted for light in the exploration of particles smaller than the wavelength of light. It is on the basis of his work that electron microscopy originated, and it has been constantly used and redefined since then.

Ryle, Sir Martin. Born: September 27, 1918, Brighton, Sussex, England; died: October 14, 1984, Cambridge, England. Specialized in radio astronomy, and is best known for the invention of a new radio telescope and the formulation of new techniques of celestial observation. Recipient of the Nobel Prize in Physics for 1974, along with Antony Hewish, for joint work with the corecipient, with whom he had collaborated for 25 years at Cavendish Laboratory, University of Cambridge. Note that radio astronomy is based upon the concept that celestial phenomena that occurred billions of years ago produced radio waves that have taken that much time to reach present detection devices. Through such radio observations, the events of the universe can be studied in a unique manner.

Salam, Abdus. Born: January 29, 1926, Jhang, India. Specialized in particle physics, and is best known for the theory relating the electromagnetic force and the weak force of the atomic nucleus. Recipient of the Nobel Prize in Physics, along with Steven Weinberg and Sheldon Glashow, for the above-mentioned theory and for discovery of the phenomenon of the "weak neural current" developed when an electron changes into a neutrino, which is in turn changed back to an electron.

Schawlow, Arthur, L. Born: May 5, 1921, Mount Vernon, New York. Specialized in optics and laser spectroscopy, and is best known for extending masers into optical applications that eventually brough the discovery of lasers. Recipient of the Nobel Prize in Physics for 1981, along with Nicolaas Bloembergen and Kai M. Siegbahn, for teamwork at Stanford University derived from the observations of the main characteristics of lasers. From this work a series of laser applications made it possible to study the properties of molecules, atoms, and nuclei with great accuracy.

Schrieffer, John Robert. Born: May 31, 1931, Oak Park, Illinois. Specialized in superconductivity, and is best known for his statistical techniques in association with superconductivity. Recipient of the Nobel Prize in Physics for 1972, along with Leon N. Cooper and John Bardeen, for his major contributions to teamwork in association with Cooper and Bardeen that paved the way for the great advances achieved in the field of superconductivity.

Schrödinger, Erwin. Born: August 12, 1887, Vienna, Austria; died: January 4, 1961, Vienna, Austria. Specialized in atomic theory and wave mechanics, and is best known for his "wave equation" concerning the mechanical properties of electrons, protons, atoms, and molecules. Recipient of the Nobel Prize in Physics for 1933, along with Paul A. M. Dirac, for the formulation of a theory of new mechanics for matter waves. The wave equations that he formulated for the motion of electrons follow the wave equation describing the propagation of light.

Schwartz, Melvin. Born: November 2, 1932, New York, New York. Specialized in high-energy particle physics, and is best known for the neutrino beam artificially produced in the laboratory, together with the cowinners of the prize. Recipient of the Nobel Prize in Physics for 1988, along with Jack Steinberger and Leon M. Lederman, for their joint work on the so-called weak interaction, defined as one of the primary forces of nature (gravitational, electromagnetic, strong, and weak). Using neutrino beams produced for the first time in the laboratory, they discovered a new type of neutrino, advancing the theory of the "standard model" in the field of particle physics.

Schwinger, Julian Seymour. Born: February 12, 1918, New York, New York. Specialized in quantum electrodynamics, and is best known for the formulation of a theory for the interaction of photons, electrons, and positrons. Recipient of the Nobel Prize in Physics for 1965, along with Richard P. Feynman and Shin'ichirō Tomonaga, for his contribution to the field of quantum electrodynamics. His major contribution was the method of "mathematical renormalization," through which he clarified erroneous deductions from Einstein's theory dealing with the action of charged subatomic particles. Specifically, he recomputed the charge of the electrons and the magnetic fields to finite values, correcting the assumptions of infinite values previously accepted.

Segrè, Emilio Gino. Born: February 1, 1905, Tivoli, Italy; died: April 22, 1989, Lafayette, California. Specialized in nuclear physics, and is best known for his discovery of the antiproton. Recipient of the Nobel Prize in Physics for 1959, along with Owen Chamberlain, for his discovery of the antiproton (the antiparticle of the proton), confirming a theory previously enunciated by Paul Dirac stating that each subatomic particle must have a corresponding particle having opposite charge and equal mass (for instance, − electrons and + positrons, − antiprotons and + protons, neutrons and antineutrons). Note that when a particle and an antiparticle eventually collide, they neutralize each other, changing their mass into kinetic energy or radiation. Further note that in his work at the University of California at Berkeley, Segrè used the Bevatron, in which protons could be accelerated up to a 6 billion electron volts.

Shockley, William. Born: February 13, 1910, London, England. Specialized in solid-state physics, and is best known for his work on semiconductors and subsequently for the discovery of the junction transistor (the type of transistor most frequently used). Recipient of the Nobel Prize in Physics for 1956, along with John Bardeen and Walter Houser Brattain. Working with the cowinners, he finally succeeded in the research on semiconductors, which act as current rectifiers that allow the flow of current in one direction while opposing it in the opposite direction.

Siegbahn, Kai M. Born: April 20, 1918, Lund, Sweden. Specialized in chemical physics, and is best known for his methods in electron spectroscopy, which are usable in chemical analysis. Recipient of the Nobel Prize in Physics for 1981, along with Nicolaas Bloembergen and Arthurr L. Schawlow, for his methodology for measuring the photoelectrons produced when electrons (called photoelectrons) are liberated from the surface of a metal by the action of high-intensity electromagnetic radiation. A practical application of the work involved the chemical analysis of metal corrosion and of catalytic reactions.

Sigbahn, Karl Manne Georg. Born: December 3, 1886, Orebro, Sweden; died: September 26, 1978, Stockholm, Sweden. Specialized in X-ray spectroscopy, and is best known for his work on the reflection and diffraction of X-rays by crystals. Recipient of the Nobel Prize in Physics for 1924 for measurements of the X-ray spectra of many elements, conducted with extreme accuracy, including the necessary methodology and instrumentation.

Stark, Johannes. Born: April 15, 1874, Schieckenhof, Bavaria, Germany; died: June 21, 1957, Traunstein, Bavaria, West Germany. Specialized in electrical conduction in gases, and is best known for his discovery of the action of strong electrical fields to split the spectral lines of elements. Recipient of the Nobel Prize in Physics for 1919 for his early prediction of the Doppler effect in canal rays (streams of positively charged ions) and for proving his prediction through experimentation in 1905.

Steinberger, Jack. Born: May 25, 1921, Bad Kissingen, Germany. Specialized in high-energy particle physics and is best known for the neutrino beam artificially produced in the laboratory, along with the cowinners of the prize. Recipient of the Nobel Prize in Physics for 1988, along with Melvin Schwartz and Leon M. Lederman, for their joint work on the so-called weak interaction, defined as one of the primary forces of nature (gravitational, electromagnetic, strong, and weak). Using neutrino beams produced for the first time in the laboratory, they discovered a new type of neutrino, advancing the theory of the "standard model" in the field of particle physics.

Stern, Otto. Born: February 17, 1988, Sohrau, Upper Silesia, Germany; died: August 17, 1969, Berkeley, California. Specialized in quantum physics, and is best known for his work on the determination of the magnetic moment of subatomic particles. Recipient of the Nobel Prize in Physics for 1943 for his contribution to the development of the molecular ray method, which eventually made possible the discovery of the proton's magnetic moment.

Tamm, Igor Yevgenyevich. Born: July 8, 1895, Vladivostok, Siberia; died: April 12, 1971, Moscow, Sovient Union. Specialized in particle physics and plasma physics, and is best known for his analysis of subatomic particles traveling at speeds higher than the speed of light (see note, below). Recipient of the Nobel Prize in Physics for 1958, along with Ilya Mikhailovich Frank and Pavel Alekseyevich Cherenkov, for formulating the theory, supported by the other two recipients, that when liquids are bombarded by gamma rays, a glowing phenomenon of light is generated by nuclear particles moving faster than light. Note that this is not contrary to Einstein's theory that the speed of light is the highest speed possible, as Einstein referred to the speed of light in a vacuum, but its value in a liquid is much lower.

Thomson, Sir George Paget. Born: May 3, 1892, Cambridge, England; died: September 10, 1975, Cambridge, England. Specialized in electron diffraction, and is best known for discovering the diffractions of beams of electrons acting as light. Recipient of the Nobel Prize in Physics for 1937, along with Clinton J. Davisson, for his experiments, independent of those of the cowinner, that supported the wave theory of matter, previously proposed by Louis de Broglie.

Dirac, Paul Adrien Maurice. Nobel Laureate in Physics 1933.

Fermi, Enrico. Nobel Laureate in Physics 1938. Copyright © The Nobel Foundation. Used with permission.

uli, Wolfgang. Nobel Laureate in Physics 1945. Copyright © The Nobel Foundation.
sed with permission.

Taylor, Richard. Born: November 2, 1929, Medicine Hat, Alberta, Canada. Specialized in nuclear physics, and is best known for his work in collaboration with the Stanford Linear Accelerator Center (SLAC), while working at Stanford University. The experimentation referred to as "SLAC-MIT" involved the use of the two-mile-long linear accelerator at Stanford, in which the structure of nucleons (protons and neutrons) was studied using record-high-energy electrons as probes. Recipient of the Nobel Prize in Physics for 1990, along with Jerome Friedman and Henry Kendall, for their joint work in pioneering investigations concerning deep inelastic scattering of electrons on protons and bound neutrons, which have been of essential importance for the development of the quark model in particle physics.

Thomson, Sir Joseph John. Born: December 18, 1856, Cheetham Hill, near Manchester, England; died: August 30, 1940, Cambridge, England. Specialized in particle physics, and is best known for the discovery of the electron, which marked the beginning of the investigation into the structure of the atom. Recipient of the Nobel Prize in Physics for 1906 for his experimental work on cathode rays, demonstrating the consistency of their particles, whose mass he was able to determine.

Ting, Samuel, C. C. Born: January 27, 1936, Ann Arbor, Michigan. Specialized in particle physics, and is best known for the discovery of a subatomic particle, the J-psi. Recipient of the Nobel Prize in Physics for 1976, along with Burton Richter, for their simultaneous but independent discovery of the J-psi particle. Referred to as the "fourth charmed quark," the J-psi particle derives its name from the combination of "J," given it by Ting, and "psi," given it by Richter.

Tomonaga, Shin'ichirō. Born: March 31, 1906, Tokyo, Japan; died: July 8, 1979, Tokyo, Japan. Specialized in quantum electrodynamics, and is best known for his completely relativistic quantum field theory. Recipient of the Nobel Prize in Physics for 1965, along with Richard Feynman and Julian Schwinger, for his independent work in Japan on quantum electrodynamics, which was in agreement with the later findings of the cowinners in the United States.

Townes, Charles Hard. Born: July 28, 1915, Greenville, South Carolina. Specialized in quantum electronics, and is best known for his invention of the maser. Recipient of the Nobel Prize in Physics for 1964, along with Nikolay Gennadiyevich Basov and Aleksandr Mikhailovich Prokhorov, for his discovery of microwave amplification by stimulated emission of radiation (named "maser" from the initials of the terms). Working in the United States independently of the Soviet cowinners, Townes arrived at his findings while studying the amplification of microwaves through atom emissions. Note that one area of application of masers is based on their ability to work as very sensitive radio receivers for short waves, and they have wide applications in radio astronomy.

Van der Meer, Simon. Born: November 24, 1925, The Hague, Netherlands. Specialized in high-energy particle physics, and is best known for the invention of the "stochastic cooling" process. Recipient of the Nobel Prize for Physics for 1984, along with Carlo Rubbia, for processes (stochastic cooling) that generated a high concentration of antiprotons to allow the collision between antiprotons and protons from which W and Z particles were produced. With his experiments the theory previously formulated by Glashow, Salam, and Weinberg was finally proved.

Van der Waals, Johannes Diderik. Born: November 23, 1837, Leiden, Netherlands; died: March 8, 1923, Amsterdam, Netherlands. Specialized in equation of state theory, and is best known for formulation of the equation of state, which explains the behavior of gases and liquids at varying temperatures and pressures. Recipient of the Nobel Prize in Physics for 1910 for his studies on the behavior of gases under high pressure near the liquefaction point and determining the discrepancy of the behavior with Boyle's law on gases under such conditions, implying a new behavioral law and the existence of a new type of molecular attraction.

Van Vleck, John H. Born: March 13, 1899, Middletown, Connecticut; died: October 27, 1980, Cambridge, Massachusetts. Specialized in magnetism, quantum mechanics, and solid-state physics, and is best known for his theories on molecular bonding and molecular spectra. Recipient of the Nobel Prize in Physics for 1977, along with Philip W. Anderson and Sir Nevill Mott, for his studies of electron motions, both rotational and translational, that relate to the magnetic properties of matter.

Walton, Ernest Thomas Sinton. Born: October 6, 1903, Dungarvan, County Waterford, Ireland. Specialized in nuclear physics, and is best known for proving the possibility of splitting the atomic nucleus by bombarding it. Recipient of the Nobel Prize in physics for 1951, along with John Cockcroft, for building an accelerator (Cockcroft-Walton accelerator) that at a voltage of 600,000 V produced beam of protons that generated two nuclei of helium from a thin film of metallic lithium.

Weinberg, Steven. Born: May 3, 1933, New York, New York. Specialized in particle physics, and is best known for the formulation of the electroweak theory. Recipient of the Nobel Prize in Physics for 1979, along with Sheldon Glashow and Abdus Salam, for his work, in combination with the other prizewinners, that resulted in the electroweak theory, which interrelates the weak force of the atomic nucleus and the electromagnetic force.

Wien, Wilhelm. Born: January 13, 1864, Gaffken, near Fischhausen, East Prussia; died: August 30, 1928, Munich, Germany. Specialized in thermal radiation, and is best known for formulation of the displacement law and the distribution law named after him. Recipient of the Nobel Prize in Physics for 1911 for devising a "black body" that is impervious to light and nonreflective and for the law relating the wavelength and the temperature of black body radiations (displacement law)—as well as for a distribution law that later proved incorrect yet led to Max Planck's theory.

Wigner, Eugene Paul. Born: November 17, 1902, Budapest, Hungary. Specialized in atomic theory, and is best known for establishing the principle of symmetry of the properties of the atomic nucleus. Recipient of the Nobel Prize in Physics for 1963, along with Maria Goeppert Mayer and J. Hans D. Jensen, for his pioneering work on the law of symmetry concerning the motions of nuclear particles and for his deductions describing the interacting force between protons and neutrons (proving that such force increases as the distance between the two particles increases, and that the force decreases as such distance decreases), as well as for his proposed models explaining the motion of the nucleons.

Wilson, Charles Thomson Rees. Born: February 14, 1869, Glencorse, Midlothian, Scotland; died: November 15, 1959, Carlops, Peeblesshire, Scotland. Specialized in ionizing particles and atmospheric electricity, and is best known for the development of the cloud chamber. Recipient of the Nobel Prize in Physics for 1927, along with Arthur Holly Compton, for developing an apparatus (cloud chamber) that made visible the paths of X rays and ionizing particles when they were illuminated, allowing them to be recorded as they were photographed. Note that the use of the cloud chamber made many more discoveries possible in later years.

Wilson, Kenneth G. Born: June 8, 1936, Waltham, Massachusetts. Specialized in elementary particle theory, and is best known for a renormalization group theory that he derived from a method used in theoretical physics, by means of which a major complex problem is eventually solved by subdividing it into a series of smaller problems that are easier to solve individually. Recipient of the Nobel Prize in Physics for 1982 for the "renormalization group theory," which he derived.

Wilson, Robert W. Born: January 10, 1936, Houston, Texas. Specialized in radio astronomy, and is best known for the discovery of the cosmic microwave background radiation accompaning the so-called big bang at the creation of the universe. Recipient of the Nobel Prize in Physics for 1978, along with Arno A. Penzias and Pytor L. Kapitsa, for his work conducted in combination with Wilson. Their work was the first to observe the residue of the radiations that accompanied the explosion at the beginning of the universe. In their experimentation on radio radiation in the galaxy, they detected an unknown radiation of equal intensity in all directions. Such radiations were confirmed by others to remnants of the original radiations that had occurred 15 billion years earlier, at the time of the explosion at the birth of the university, now cooled down enough to be detectable as radio waves.

Yang, Chen Ning. Born: September 22, 1922, Hogei, Anhwei, China. Specialized in particle physics and statistical mechanics, and is best known for his suggestions, together with those of the cowinner of the prize, concerning theory and experimentation proving the absence of conservation of parity. Recipient of the Nobel Prize in Physics for 1957, along with Tsung-Dao Lee, for their joint work on the theoretical aspects of the law of conservation of parity, which led to later discoveries proving that parity was not conserved.

Yukawa, Hideki. Born: January 23, 1907, Tokoyo, Japan; died: September 8, 1981, Kyoto, Japan. Specialized in nuclear physics, and is best known for his work predicting the existence of new particles, "mesons," which were later discovered. Recipient of the Nobel Prize in Physics for 1949 for his theory on the force of attractions between protons and neutrons in the nucleus and for proposing the existence of "mesons," field particles that he predicted could also be found outside the nucleus in cosmic radiations.

Zeeman, Pieter. Born: May 25, 1865, Isle of Schouwen, Zeeland, Netherlands; died: October 9, 1943, Amsterdam, Netherlands. Specialized in electromagnetic theory and magneto-optics, and is best known for demonstrating that electrical wave-motion and light were exactly the same in nature. Recipient of the Nobel Prize in Physics for 1902, along with Hendrick Antoon Lorentz, for providing an experimental basis for Lorentz's "electron theory."

Zernike, Frits. Born: July 16, 1888, Amsterdam, Netherlands; died: March 10, 1966, Groningen, Netherlands. Specialized in optics, and is best known for the phase contrast method and for the phase contrast microscope. Recipient of the Nobel Prize in Physics for 1953 for discovery of the theory of the phase contrast method and for invention of the phase contrast microscope, which enabled accurate measurement of particles with dimensions smaller than that of the wavelength of light, to the point that atomic structure eventually became visible.

Chronological List of Nobel Prize Winners in Physics, 1901–1989

This list is included to show the progressive development of the physical sciences in the twentieth century, as indicated by evaluations of the work of the world's most prominent scientists by representatives of the Nobel Foundation.*

1901	Wilhelm Conrad Röntgen (1845–1923)	Germany	X-radiation
1902	Hendrik Antoon Lorentz (1853–1923)	Netherlands	electromagnetic theory
	Pieter Zeeman (1865–1943)	Netherlands	electromagnetic theory/ magnetooptics
1903	Antoine-Henri Becquerel (1852–1908)	France	radioactivity
	Pierre Curie (1859–1906)	France	radioactivity/ magnetism/ crystallography
	Marie Curie (1867–1934)	Poland/France	radioactivity
1904	Lord Rayleigh (1842–1919)	Great Britain	acoustics/optics
1905	Philipp Lenard (1862–1947)	Germany	photoelectricity/ electrons
1906	Sir Josephson John Thomson (1856–1940)	Great Britain	particle physics
1907	Albert Abraham Michelson (1852–1931)	United States	optics/spectroscopy/ interferometry
1908	Gabriel Lippmann (1845–1921)	France	applied mathematical physics
1909	Guglielmo Marconi (1874–1937)	Italy	radio telegraphy
	Karl Ferdinand Braun (1850–1918)	Germany	wireless telegraphy

No awards were made in 1916, 1931, 1934, and 1940–1942.

1910	Johannes Diderik van der Waals (1837–1923)	Netherlands	equation of state theory
1911	Wilhelm Wien (1864–1928)	Germany	thermal radiation
1912	Nils Gustaf Dalén (1869–1937)	Sweden	engineering
1913	Heike Kamerlingh Onnes (1853–1926)	Netherlands	low-temperature physics

CRITICAL PERIOD: FIRST WORLD WAR

1914	Max von Laue (1879–1960)	Germany	X-ray optics
1915	Sir William Henry Bragg (1862–1942)	Great Britain	radioactivity/X-ray spectroscopy/X-ray crystallography
	Sir Lawrence Bragg (1890–1971)	Great Britain	X-ray crystallography
1916	Reserved	—	—
1917	Charles Glover Barkla (1877–1944)	Great Britain	X-radiation/ secondary radiation
1918	Max Planck (1858–1947)	Germany	quantum physics

1919	Johannes Stark (1874–1957)	Germany	electrical conduction in gases
1920	Charles-Édouard Guillaume (1861–1938)	Switzerland	metallurgy/metrology
1921	Albert Einstein (1879–1955)	Germany/ Switzerland/ United States	theoretical physics
1922	Niels Bohr (1885–1962)	Denmark	atomic structure/ quantum theory
1923	Robert Andrews Millikan (1868–1953)	United States	electronic charge/ photoelectric effect
1924	Karl Manne Georg Siegbahn (1886–1978)	Sweden	X-ray spectroscopy

1925	James Franck (1882–1964)	Germany	atomic physics/ molecular physics
	Gustav Hertz (1887–1975)	Germany	atomic physics/ molecular physics
1926	Jean-Baptiste Perrin (1870–1942)	France	molecular physics
1927	Arthur Holly Compton (1892–1962)	United States	X-radiation/optics
	Charles Thomson Rees Wilson (1869–1959)	Great Britain	ionizing particles/ atmospheric electricity
1928	Sir Owen Willans Richardson (1879–1959)	Great Britain	thermionics
1929	Louis de Broglie (1892–1987)	France	quantum physics/ wave mechanics
1930	Sir Chandrasekhara Venkata Raman (1888–1970)	India	optics
1931	Reserved	—	—
1932	Werner Heisenberg (1901–1976)	Germany	quantum mechanics
1933	Erwin Schrödinger (1887–1961)	Austria	atomic theory/wave mechanics
	Paul Adrien Maurice Dirac (1902–1984)	Great Britain	quantum mechanics
1934	Reserved	—	—
1935	Sir James Chadwick (1891–1974)	Great Britain	atomic physics/ nuclear physics
1936	Victor Franz Hess (1883–1964)	Austria	cosmic radiation
	Carl David Anderson (1905–)	United States	particle physics
1937	Clinton Joseph Davisson (1881–1958)	United States	electron physics
	Sir George Paget Thomson (1892–1975)	Great Britain	electron diffraction

1938	Enrico Fermi (1901–1954)	Italy	radioactivity/nuclear reactions

CRITICAL PERIOD: SECOND WORLD WAR

1939	Ernest Orlando Lawrence (1901–1958)	United States	nuclear physics
1940–1942	Reserved	—	—
1943	Otto Stern (1888–1969)	United States	quantum physics
1944	Isidor Isaac Rabi (1898–1988)	United States	nuclear physics
1945	Wolfgang Pauli (1900–1958)	Austria/United States	quantum mechanics

1946	Percy Williams Bridgman (1882–1961)	United States	high-pressure physics
1947	Sir Edward Victor Appleton (1892–1965)	Great Britain	radio physics/atmospheric physics
1948	Patrick M. S. Blackett (1897–1974)	Great Britain	nuclear physics/cosmic radiation
1949	Hideki Yukawa (1907–1981)	Japan	nuclear physics
1950	Cecil Frank Powell (1903–1969)	Great Britain	nuclear physics/cosmic radiation
1951	Sir John Douglas Cockcroft (1897–1967)	Great Britain	nuclear physics
	Ernest Thomas Sinton Walton (1903–)	Ireland	nuclear physics
1952	Felix Bloch (1905–1983)	United States	nuclear physics
	Edward Mills Purcell (1912–)	United States	nuclear magnetic resonance

1953	Frits Zernike (1888–1966)	Netherlands	optics
1954	Max Born (1882–1970)	Great Britain	quantum mechanics
	Walther Bothe (1891–1957)	West Germany	particle physics/ nuclear energy
1955	Willis Eugene Lamb, Jr. (1913–)	United States	quantum electrodynamics
	Polykarp Kusch (1911–)	United States	atomic physics/ molecular physics
1956	William Shockley (1910–)	United States	solid-state physics
	John Bardeen (1908–)	United States	solid-state physics
	Walter H. Brattain (1902–1987)	United States	solid-state physics
1957	Chen Ning Yang (1922–)	China/United States	particle physics/ statistical mechanics
	Tsung-Dao Lee (1926–)	China/United States	particle physics/ statistical mechanics
1958	Pavel Alekseyevich Cherenkov (1904–)	Soviet Union	nuclear physics/ particle physics
	Ilya Mikhailovich Frank (1908–)	Soviet Union	nuclear physics/ particle physics/ optics
	Igor Yevgenyevich Tamm (1895–1971)	Soviet Union	particle physics/ plasma physics
1959	Emilio Gino Segrè (1905–)	United States	nuclear physics
	Owen Chamberlain (1920–)	United States	nuclear physics
1960	Donald A. Glaser (1926–)	United States	particle physics
1961	Robert Hofstadter (1915–)	United States	nuclear physics

ll-Mann, Murray. Nobel Laureate in Physics 1969. Copyright © The Nobel Foundation.
ed with permission.

Penzias, Arno A. Physics 1978. Copyright © The Nobel Foundation. Used with permission

Vilson, Robert W. Nobel Prize for Physics, 1978. Copyright © The Nobel Foundation. Used with permission.

Weinberg, Steven, Nobel Laureate in Physics, 1979. Copyright © The Nobel Foundation. Used with permission.

	Rudolf Ludwig Mössbauer (1929–)	West Germany	gamma radiation
1962	Lev Davidovich Landau (1908–1968)	Soviet Union	quantum mechanics
1963	Eugene Paul Wigner (1902–)	Hungary/ United States	atomic theory
	Maria Goeppert Mayer (1906–1972)	Germany/ United States	nuclear physics
	J. Hans D. Jensen (1907–1973)	West Germany	nuclear physics
1964	Charles Hard Townes (1915–)	United States	quantum electronics
	Nikolay Gennadiyevich Basov (1922–)	Soviet Union	quantum electronics
	Aleksandr Mikhailovich Prokhorov (1916–)	Soviet Union	quantum radiophysics/ quantum electronics
1965	Shin'ichirō Tomonaga (1906–1979)	Japan	quantum electrodynamics
	Julian Seymour Schwinger (1918–)	United States	quantum electrodynamics
	Richard P. Feynman (1918–1988)	United States	quantum electrodynamics
1966	Afred Kastler (1902–1984)	France	optical spectroscopy/ Hertzian resonances
1967	Hans Albrecht Bethe (1906–)	United States	nuclear physics/ astrophysics
1968	Luis W. Alvarez (1911–1988)	United States	high-energy particle physics
1969	Murray Gell-Mann (1929–)	United States	particle physics

1970	Hannes Alfvén (1908–)	Sweden	plasma physics
	Louis-Eugène-Félix Néel (1904–)	France	nuclear magnetism
1971	Dennis Gabor (1900–1979)	Great Britain	electron optics/ holography
1972	John Bardeen (1908–)	United States	superconductivity
	Leon N Cooper (1930–)	United States	superconductivity
	John Robert Schrieffer (1931–)	United States	superconductivity
1973	Leo Esaki (1925–)	Japan	quantum mechanics/ solid-state physics
	Ivar Giaever (1929–)	Norway/United States	quantum mechanics/ solid-state physics/ biophysics
	Brian D. Josephson (1940–)	Great Britain	quantum mechanics/ solid-state physics
1974	Sir Martin Ryle (1918–1984)	Great Britain	radio astronomy
	Antony Hewish (1924–)	Great Britain	radio astronomy
1975	Aage Bohr (1922–)	Denmark	nuclear physics
	Ben R. Mottelson (1926–)	Denmark	nuclear physics
	L. James Rainwater (1917–1986)	United States	structural nuclear physics
1976	Burton Richter (1931–)	United States	particle physics
	Samuel C. C. Ting (1936–)	United States	particle physics
1977	John H. Van Vleck (1899–1980)	United States	magnetism/quantum mechanics/solid-state physics
	Sir Nevill Mott (1905–)	Great Britain	solid-state physics
	Philip W. Anderson (1923–)	United States	solid-state physics

1978	Pyotr Leonidovich Kapitsa (1894–1984)	Soviet Union	low-temperature physics/plasma physics
	Arno A. Penzias (1933–)	Germany/ United States	radio astronomy
	Robert W. Wilson (1936–)	United States	radio astronomy
1979	Sheldon L. Glashow (1932–)	United States	particle physics
	Abdus Salam (1926–)	Pakistan	particle physics
	Steven Weinberg (1933–)	United States	particle physics
1980	James W. Cronin (1931–)	United States	particle physics
	Val. L. Fitch (1923–)	United States	particle physics
1981	Nicolaas Bloembergen (1920–)	United States	optics/quantum electronics
	Arthur L. Schawlow (1921–)	United States	optics/laser spectroscopy
	Kai M. Siegbahn (1918–)	Sweden	chemical physics
1982	Kenneth G. Wilson (1936–)	United States	elementary particle theory
1983	Subrahmanyan Chandrasekhar (1910–)	United States	astrophysics
	William A. Fowler (1911–)	United States	astrophysics/nuclear physics
1984	Carlo Rubbia (1934–)	Italy	high-energy particle physics
	Simon van der Meer (1925–)	Netherlands	high-energy particle physics
1985	Klaus von Klitzing (1943–)	West Germany	condensed-matter physics
1986	Ernst Ruska (1906–1988)	West Germany	electrical engineering/ electron microscopy

	Gerd Binnig (1947–)	West Germany	scanning tunneling microscopy
	Heinrich Rohrer (1933–)	Switzerland	scanning tunneling microscopy
1987	Karl Alexander Müller (1927–)	Switzerland	solid-state physics/ superconductivity
	J. Georg Bednorz (1950–)	West Germany	solid-state physics/ superconductivity
1988	Leon M. Lederman (1922–)	United States	high-energy particle physics
	Melvin Schwartz (1932–)	United States	high-energy particle physics
	Jack Steinberger (1921–)	Germany/ United States	high-energy particle physics
1989	Hans G. Dehmelt (1922–)	Germany	
	Wolfgang Paul (1913–)	Germany	
	Norman F. Ramsey (1915–)	United States	

4
Scientists and Units

Many scientists contributed to the body of knowledge that led to present-day physics, from its origins in the Greek school of philosophy up through the twentieth century. The names of the vast majority are strictly connected to their scientific contributions, but a few of them have been specially recognized by having their names associated with the units of measure that they originated. Brief biographical sketches follow of scientists whose names have been used for units in physics.

Ampère, André Marie (1775–1836). French physicist, a professor at the Ecole Polytechnique, Paris, and major developer of electrodynamics. Named after him is the ampere, the unit measuring electric current flowing through a conductor having a resistance of one ohm under an electromotive force of one volt, or defined as the electric flow of one coulomb of electric charge per second through a conductor.

Ångström, Anders Jonas (1814–1874). Swedish physicist, a contributor in optics and spectroscopics. Named after him is the angström, a unit of length equal to 10^{-10} meter, used for measuring the wavelength of light and other small dimensions.

Bell, Alexander Graham (1847–1922). English-born American scientist who invented the telephone in 1876. Named after him is the bel (10 decibels), a unit of intensity of sound, measuring the amount of sound energy that is transmitted to one square centimeter of the ear.

Celsius, Anders (1701–1744). Born in Uppsala, Sweden, Celsius, was the originator of the first centigrade scale in 1742. Named after him is the scale and the unit of temperature still in use.

Coulomb, Charles-Augustin de (1736–1806). French scientist, the discoverer of Coulomb's law, whose name is given to a unit of electric charge in MKS and SI systems, defined as the quantity of charge that passes through a cross section of a conductor in one second when the current equals one ampere.

Curie, Pierre (1859–1906). French Physicist, remembered in the history of science in in association with his wife (Marie Sklodowska) for their teamwork that led to the discovery of radium and polonium. After they jointly received the Nobel Prize in physics in 1903, Marie attained a second Nobel Prize in 1911 in chemistry. A graduate of the Sorbonne in Paris, Dr. Pierre Curie spent most of his life as a professor at the School of Industrial Physics and Chemistry of Paris and was later appointed professor of general physics at the Sorbonne, where he remained until his death. Among his achievements are his early work on the electric and magnetic properties of crystals, the piezoelectric effect, and the Curie point (critical temperature at which ferromagnetic materials lose most of their magnetism). Named after him is the Curie, a unit of physics that measured the amount of radon (a gaseous radioactive element derived from the disintegration of radium) emanated by one gram of radium, which was redefined in 1953 as the quantity of radionuclide in which the number of disintegrations per second is 3.000×10^{10}.

Dalton, John (1766–1844). English chemist and physicist who was the formulator of the atomic theory and famous for the Dalton law relating the pressures of the individual components in a mixture of gases. Named for him is the dalton, a unit of atomic mass.

Fahrenheit, Daniel Gabriel (1686–1736). Born in Danzig, Poland (annexed to Germany shortly later), Fahrenheit was the originator of the temperature scale named after him. Also the basic unit of temperature within that scale carries his name.

Faraday, Michael (1791–1867). English scientist who pioneered in the field of electromagnetic induction and was the father of the electric motor. The two basic laws of electrolysis carry his name, as well as the Faraday unit, which measures the electric charge necessary to fill one gram-equivalent of a substance by electrolysis.

Gauss, Carl Friedrich (1777–1885). German mathematician, one of the greatest of all time, who contributed to the development of non-Euclidean geometry and formulated the so-called Gaussian curvature. He contributed also to geodesy, magnetism, electricity, and statistics. Named for him is the gauss, the unit of magnetic induction equal to one Maxwell per square centimeter or 10^{-4} weber per square meter.

Gilbert, William (1540–1603). English physician and physicist who was the first European to accurately describe the magnetism of the earth and the behavior of magnets. From his studies of electricity and magnetism were derived the terms "electricity," "electric force," and "magnetic pole." A unit of magnetic force, the gilbert, is named after him. He wrote *De Magnete* about 1600.

Giorgi, Giovanni (1871–1950). Professor of engineering at the University of Rome who worked in hydroelectric installations, electric distribution networks, and electric transportation systems. Inventor of the system of units, including mechanical, electrical, and magnetic units of measure, called the Giorgi International System of Measurement (MKSA System) in 1901, and endorsed in 1960 by the General Conference of Weights and Measures.

Henry, Joseph (1797–1878). American physicist. Inventor of the unit of inductance that was named after him.

Joule, James Prescott (1818–1889). English physicist. Developer of the first law of thermodynamics, enunciating the conservation of energy. Named after him is the joule, a unit of work and energy in the mks system, which measures the work produced by a force of one newton when its point of application moves one meter in the direction of the force.

Lambert, Johann Heinrich (1728–1777). German scientist and philosopher. Named after him is the unit of measure of the intensity of light, the lambert.

Mach, Ernst (1838–1916). German physicist, for whom the Mach number was named. Such a number is defined as the ratio between the speed of a body or fluid and the speed of sound in a medium. Thus, establishing the speed of sound in a medium.

So, multiplying the mach number by the speed of sound in a medium, the Mach becomes a unit of velocity.

Maxwell, James Clerk (1831–1879). Scottish mathematical physicist. Formulator of the general equation of the electromagnetic field, who extended the Faraday concept in electromagnetism, established the electromagnetic nature of light, and made a significant contribution to the development of the kinetic theory of gases. He contributed substantially to elastic theory in structural engineering. As a professor at King's College in London, his scientific contribution gave him a prominent role in nineteenth-century physics. The unit named after him is a measure of magnetic flux.

Newton, Sir Isaac (1642–1727). English physicist, the most outstanding figure in the world of science. His major work included laws governing motion and gravitational attraction involving planetary motion, as well as exact measurement of the masses of the sun and planets, determination of the path of comets, and the relationship between tides and lunar attaction. His work in optics focused on the composition of white light. His astonishing contribution to mathematics produced infinitesimal calculus, with the two divisions of differential and integral calculus. Named after him is the Newton, a unit of force in the MKS system, measuring the force that would give one kilogram of mass an acceleration of one meter/second^2.

Ohm, Georg Simon (1787–1854). German mathematician and physicist and professor at Munich, who developed the relationship between the basic parameters controlling electrical currents. Named after him is the ohm, a unit measuring the electrical resistance of a conductor that carries a current of one ampere under an electromotive force of one volt.

Planck, Max Karl Ernst Ludwig (1858–1947). German physicist and professor at Kiel and Berlin who was the father of the quantum theory, which states that energy is not absorbed or radiated continuously, but is radiated discontinuously in definite units called quanta. The Planck constant (*h*), which Planck called the "quantum of action," is a universal constant that when multiplied by the frequency of the oscillating particles of a black body gives the element of energy of the oscillator.

Réaumur, René-Antoine (1683–1757). Born in La Rochelle, France, Réaumur, a scientist in entomology, was the originator of a temperature scale named after him. Carrying his name is also the unit of temperature included in such a scale.

Stokes, Sir George Gabriel (1819–1903). British mathematician and physicist who was a major contributor to hydrodynamics, particularly in the area of viscosity. Stokes' law determines the velosity of a sphere falling within a viscous fluid. Named after him is the stoke, a unit of kinematic viscosity measuring the viscosity of a fluid with a dynamic viscosity of one poise and a density of one gram per cubic centimeter.

Thomson, Sir William (later Lord Kelvin) (1824–1907). Born in Belfast, Ireland, Lord Kelvin was the originator of the absolute temperature scale that incorporated the celsius scale within it. The scale and the unit temperature Kelvin that are still in use are named after him.

Torricelli, Evangelista (1608–1647). Italian mathematician and physicist, assistant to Galileo, and professor at the Florentine Academy, who is famous for the determination of atmospheric pressure, the invention of the barometer, and the "Torricelli theorem" in hydrodynamics. Named after him is the torr, a unit of pressure, practically equal to 1 mm of Hg.

Volta, Alessandro, Conte (1745–1827). Italian physicist and professor at Pavia who is particularly famous for his works in electricity. He first generated the condition for the flow of electricity in a circuit by means of an electromotive force attained through a device named after him (voltaic pile). The volt, a unit of electromotive force, also was named for him; it measures the potential difference of an electromotive force that generates a current of one ampere in a circuit having a resistance of one ohm. The volt also can be defined as the potential difference between two points in a conductor such that it will produce a flow of one coulomb generating one joule of work.

Watt, James (1736–1819). Scottish scientific-instrument maker and major contributor to the improvement of the steam engine. Named after him is the watt, a unit of electric power in the mks system generated by a current of one ampere under a potential differential of one volt. One watt is equal to one joule per second.

Weber, Wilhelm Edward, (1804–1891). German physicist and professor at Göttingen, where he and Gauss organized the Göttingen Magnetic Union for worldwide study of terrestrial magnetism. Named after him is the weber, a unit of magnetic flux in electromotive force that generates a current of one ampere in a circuit having a resistance of one ohm. It also can be defined as the potential difference between two points in a conductor such that it will produce a flow of one coulomb, generating one joule of work.

5
Units in the Physical Sciences and Technology

The quantities addressed in this chapter are part of the terminology of the physical sciences and related technological fields that is now in use. The list of these quantities presented here includes units of measure and their symbols, as well as the systems of which they are a part, at times specifying the major countries in which they have been adopted. Some quantities may not have been included, especially if considered obsolete, but the list that follows attempts to be as complete as is practicable.

abampere

symbol	aA
system	electromagnetic CGS
classification	unit of electric current

abampere centimeter squared

symbol	$aA \cdot cm^2$
system	electromagnetic CGS
classification	unit of electromagnetic moment

abampere per square centimeter

symbol	aA/cm^2
system	electromagnetic CGS
classification	unit of current density

abcoulomb

symbol	aC
system	electromagnetic CGS
classification	unit of electric charge

abcoulomb centimeter

symbol	$aC \cdot cm$
system	electromagnetic CGS
classification	unit of electric dipole moment

abcoulomb per cubic centimeter

symbol	aC/cm^3
system	electromagnetic CGS
classification	unit of volume density of electric charge

abcoulomb per square centimeter

symbol	aC/cm^2
system	electromagnetic CGS
classification	unit of electric flux density and unit of electric polarization

abfarad

symbol	aF
system	electromagnetic CGS
classification	unit of capacitance

abhenry

symbol	aH
system	electromagnetic CGS
classification	unit of inductance

abmho

symbol	a℧
(*see* absiemens)	

abohm

symbol	aΩ
system	electromagnetic CGS
classification	unit of resistance

abohm centimeter

symbol	aΩ · cm
system	electromagnetic CGS
classification	unit of resistivity

absiemens

symbol	aS
system	electromagnetic CGS
classification	unit of conductance

absiemens per centimeter

symbol	aS/cm
system	electromagnetic CGS
classification	unit of conductivity

abtesla

symbol	aT
(*see* gauss)	

abvolt

symbol	aV
system	electromagnetic CGS
classification	unit of electric potential

abvolt per centimeter

symbol	aV/cm
system	electromagnetic CGS
classification	unit of strength of electric field

abweber

symbol	aWb
(*see* maxwell)	

acoustic ohm (not in use)

classification	unit of impedance in acoustics

acre

classification	unit of area, used in land surveying
system	imperial unit
country	United States, United Kingdom

acre-foot

symbol	acre · ft
classification	unit of volume
country	United States

acre-foot per day

symbol	acre · ft/d
classification	unit of flow rate of volume
country	United States

acre-foot per hour

symbol	acre · ft/h
classification	unit of flow rate of volume
country	United States

acre-inch

symbol	acre-inch
clssification	unit of volume
country	United States

acre per pound

symbol	acre/lb
classification	unit of specific surface
country	United States, United Kingdom

ampere

symbol A
system SI (base unit)
classification unit of electric current, current linkage, magnetic potential difference, and magnetomotive force

ampere-circular mil

symbol A-circular mil
classification unit of electromagnetic moment
country United States, United Kingdom

ampere hour

symbol A · h
system non-SI (approved)
classification unit of electric charge

ampere meter squared

symbol $A \cdot m^2$
system SI
classificaton unit of electromagnetic moment

ampere minute

symbol A · min
system non-SI (approved)
classification unit of electric charge

ampere per inch

symbol A/in
classification unit of field of magnetic strength
country United States, United Kingdom

ampere per kilogram

symbol A/kg
(*see* coulcomb per kilogram second)

ampere per meter

symbol	A/m
system	SI
classification	unit of strength of magnetic field, linear current density, and magnetization

ampere per square inch

symbol	A/in^2
classification	unit of current density

ampere per square meter

symbol	A/m^2
classificaton	unit of current density

ampere per square meter kelvin squared

symbol	$A/(m^2 \cdot K^2)$
system	SI
classification	unit of Richardson constant

ampere per volt

symbol A/V
(*see* siemens)

ampere per weber

symbol A/Wb
(*see* reciprocal henry)

ampere second

symbol $A \cdot S$
(*see* coulomb)

ampere square meter

symbol	$A \cdot m^2$
system	SI
classification	unit of magnetic moment, Bohr magneton, and nuclear magneton

ampere square meter per joule second

symbol $A \cdot m^2/(J \cdot s)$
system SI
classification unit of gyromagnetic coefficient

ampere-turn (not in use)

symbol At
classification unit of magnetomotive force

ampere-turn per meter

symbol At/m
(*see* ampere per meter)

ångström

symbol Å
classification unit of wavelength

api

symbol A/in
classification ampere per inch

apostilb

symbol asb
classification unit of luminance

ara

symbol a
classification unit of area, used in land surveying, equal to 100 m^2

assay ton

classification unit of mass (32.667 g)
country United Kingdom

assay ton

classification unit of mass (29.167 g)
country United States

astronomical unit

symbol	AU
system	unit adopted in 1979, non-SI (approved)
classification	unit of length equal to $1.49597870 \times 10^{11}$ m

atmosphere (standard)

symbol	atm
classification	unit of pressure equal to 1.03323 kgf/cm^2

atmosphere (technical)

symbol	at
classification	unit of pressure equal to 1.0 kgf/cm^2

atomic mass unit (unified)

symbol	u
system	non-SI (approved)
classification	unit of atomic mass constant that replaces the old chemical unit and the old physical unit

bar

symbol	bar
classification	unit of pressure for fluids equal to 1.01972 kgf/cm^2

barn

symbol	b
classification	unit of area, used for cross sections and equal to $10^{-28} m^2$

barn per electronvolt

symbol	b/eV
classification	unit of spectral cross section

barn per erg

symbol	b/erg
classification	unit of spectral cross section

barn per steradian

symbol b/sr
classification unit of angular cross section

barn per steradian electronvolt

symbol b/(sr · erg)
classification unit of spectral angular cross section

barrel

classification unit of volume, used particularly for petroleum products, etc.

barye

symbol ba
(*see* dyne per square centimeter)

becquerel

symbol Bq
system SI (additional unit)
classification unit of activity of radionuclide

becquerel per cubic meter

symbol Bq/m^3
system SI
classification unit of volume activity of radionuclide

bequerel per kilogram

symbol Bq/kg
system SI
classification unit of linear activity of radionuclide

becquerel per mole

symbol Bq/mol
system SI
classification unit of molar activity of radionuclide

bel

classification multiple of the decibel
(*see* decibel

biot

symbol Bi
system CGS
classification unit of electric current equal to 0.1 ampere

biot centimeter squared

symbol $Bi \cdot cm^2$
system CGS
classification unit of electromagnetic moment

biot per centimeter

symbol Bi/cm
system CGS
classification unit of strength of magnetic field

biot second

symbol $Bi \cdot s$
system CGS
classification unit of electric charge

bit

classification binary unit of information, measuring the capacity of a bank system to store data

bit per centimeter

symbol bit/cm
classification unit of bit density (linear)

bit per inch

symbol bit/in
classification unit of bit density (linear)

bit per second

symbol bit/s
classification unit of bit rate
(*see* bit)

bit per square centimeter

symbol bit/cm^2
classification unit of bit density (surface)

bit per square inch

symbol bit/in^2
classification unit of bit density (surface)

bit per square millimeter

symbol bit/mm^2
classification unit of bit density (surface)

bit per centimeter

symbol bit/cm
classification unit of bit density (linear)

blondel

(*see* apostilb)

board foot

classification unit of volume, used for wood products, equal to 12 in × 12 in × 1 in
country United States

bougie nouvelle (not in use)

classification unit of luminous intensity, substituted for by the candela in 1948

brake horse-power

(*see* horsepower)

brewster

symbol	B
classification	unit of stress optical coefficient

British thermal unit

symbol	Btu
classification	unit of heat quantity adopted in 1956
country	United States, United Kingdom

British thermal unit foot per square foot hour degree Fahrenheit or Rankine

symbol	$Btu \cdot ft/(ft^2 \cdot h \cdot °F)$ or $Btu\ ft/(ft^2 \cdot h \cdot °R)$
classification	unit of thermal conductivity

British thermal unit inch per square foot hour degree of Fahrenheit or Rankine

symbol	$Btu\ in/(ft^2 \cdot h \cdot °F)$ or $Btu\ in/(ft^2 \cdot h \cdot °R)$
classification	unit of thermal conductivity

British thermal unit per cubic foot

symbol	Btu/ft^3
classification	unit of heat per unit of volume

British thermal unit per cubic foot hour

symbol	$Btu/(ft^3 \cdot h)$
classification	unit of heat rate

British thermal unit per foot hour degree Fahrenheit or Rankine

symbol	$Btu/(ft \cdot h \cdot °F)$ or $Btu/(ft \cdot h \cdot °R)$
classification	unit of thermal conductivity

British thermal unit per foot second degree Fahrenheit or Rankine

symbol	$Btu/(ft \cdot s \cdot °F)$ or $Btu/(ft \cdot s\ °R)$
classification	unit of thermal conductivity

Bristish thermal unit per hour

symbol Btu/h
classification unit of rate of heat flow

British thermal unit per pound

symbol Btu/lb
classification unit of heat per unit weight

British thermal unit per pound degree Fahrenheit or Rankine

symbol Btu/(lb · °F) or Btu/(lb · R)
classification unit of specific heat capacity

British thermal unit per square foot hour

symbol $\text{Btu}/(\text{ft}^2 \cdot \text{h})$
classification unit of density of heat flow rate

British thermal unit per square foot hour degree Fahrenheit or Rankine

symbol $\text{Btu}/(\text{ft}^2 \cdot \text{h} \cdot \text{F})$ or $\text{Btu}/(\text{ft}^2 \cdot \text{h} \cdot {}^\circ\text{R})$
classification unit of coefficient of heat tranfer

British thermal unit per square foot second degree Fahrenheit or Rankine

symbol $\text{Btu}/(\text{ft}^2 \cdot \text{s} \cdot {}^\circ\text{F})$ or $\text{Btu}/(\text{ft}^2 \cdot \text{s} \cdot {}^\circ\ \text{R})$
classification unit of coefficient of heat transfer

bushel

classification unit of volume equal to $3.63687 \times 10^{-2}\ \text{m}^3$
country United Kingdom

bushel

classification unit of volume for dry goods equal to $3.52391 \times 10^{-2}\ \text{m}^3$
country United States

byte

classification — unit equal to eight bits, which are binary units of information, measuring the capacity of a bank system to store data

calorie, defined

(*see* calorie, thermochemical)

calorie (dietetic)

classification — unit equal to 10^3 cal_{15}

calorie (I.T.)

symbol — cal_{IT} or cal
classification — unit of heat changed in 1956 to the international table calorie (cal_{IT})

calorie, large

(*see* kilocalorie)

calorie, mean

classification — unit equal to 4.1900 joules

calorie, small

(*see* gram-calorie)

calorie, thermochemical

symbol — cal_{th}
classification — unit equal to 4.184 joules

calorie, water

(*see* calorie, 15°C)

calorie, 15°C

symbol — cal_{15}
system — adopted in 1950 by Comite International des Poids et Mesures
classification — unit equal to 4.1855 joules

calorie, 15°C

symbol cal_{15}
system adopted in 1939 by National Bureau of Standards
classification unit equal to 4.1858 joules

calorie, 20°C

symbol cal_{20}
classification unit equal to 4.1819 joules

calorie (I.T.) per centimeter second kelvin or degree Celsius

symbol $cal_{IT}/(cm \cdot s \cdot K)$ or $cal_{IT}/(cm \cdot s \cdot °C)$
classification unit of thermal conductivity

calorie (I.T.) per gram

symbol cal_{IT}/g
classification unit of specific internal energy

calorie (I.T.) per gram kelvin or degree Celsius

symbol $cal_{IT}/(g \cdot K)$ or $cal_{IT}/(g \cdot °C)$
classification unit of specific heat capacity and specific entropy

Calorie (I.T.) per kelvin or degree Celsius

symbol cal_{IT}/K or $cal_{IT}/°C$
classification unit of heat capacity

calorie (i. T. per second)

symbol cal_{IT}/s
classification unit of rate of heat flow

calorie (I.T.) per second centimeter kelvin or degree celsius

symbol $cal_{IT}/(s \cdot cm \cdot K)$ or $cal_{IT}/(s \cdot cm \cdot °C)$
classification unit of thermal conductivity

calorie (I.T.) per second square centimeter kelvin or degree Celsius

symbol $cal_{IT}/(s \cdot cm^2 \cdot K)$ or $cal_{IT}/(s \cdot cm^2 \cdot °C)$
classification unit of coefficient of heat transfer

calorie (I.T.) per square centimeter second

symbol $cal_{IT}/(cm^2 \cdot s)$
classification unit of density of rate of heat flow

calorie (I.T.) per square centimeter second kelvin or degree Celsius

symbol $cal_{IT}/(cm^2 \cdot s \cdot K)$ or $cal_{IT}/(cm^2 \cdot s \cdot °C)$
classification unit of coefficient of heat transfer

candela

symbol cd
system SI (base unit)
classification unit of luminous intensity

candela per square centimeter

symbol cd/cm^2
system SI (multiple unit)
classification unit of luminance

candela per square foot

symbol cd/ft^2
classification unit of luminance

candela per square inch

symbol cd/in^2
classification unit of luminance

candela per square meter

symbol cd/m^2
system SI
classification unit of luminance

candle; new candle (not in use)
(*see* candela)

carat

symbol C
classification unit measuring the composition of gold

carcel (not in use)

classification unit of luminous intensity

cent

classification unit of frequency interval and reactivity (dimensionless quantities)

cental

symbol ctl
system imperial unit and avoirdupois unit
classification unit mass of equal to 4.53592×10 kg and 10^2 lb
country United Kingdom

centesimal minute

symbol . . . ^{cg}
classification unit of plane angle

centesimal second

symbol . . . ^{cc}
classification unit of plane angle that is one hundredth of a centesimal minute

centiare

symbol ca
classification unit of area equal to 1 m^2

Centigrade heat unit

symbol CHU
classification unit of heat equal to 1.8 British thermal units

centimeter

symbol cm
system SI (multiple unit) and CGS (base unit)
classification unit of length

centimeter per second squared

symbol	cm/s^2
system	CGS
classification	unit of acceleration

centimeter second degree Celsius per calorie (I.T.)

symbol	$cm \cdot s \cdot C/cal_{IT}$
classification	unit of thermal resistivity

centimeter squared per second

symbol	cm^2/s

(*see* stokes)

centipoise

symbol	cP
classification	unit of dynamic viscosity

centistokes

symbol	cSt
classification	unit of kinematic viscosity

chain

system	imperial unit
classification	unit of length equal to 2.01168×10 m
country	United States, United Kingdom

cheval vapeur

(*see* horsepower (metric))

circular inch

classification	unit of area equal to 5.06707×10^{-4} m^2 and 7.85398×10^{-1} in^2
country	United States, United Kingdom

circular mil

classification	unit of area equal to 5.06707×10^{-10} m^2 and 7.85398×10^{-7} in^2
country	United States, United Kingdom

clausius (not in use)

classification unit of entropy

clusec

classification unit of fluid escape rate, used in associated with vacuum measurements

cord

classification unit of volume, used for measuring wood, equal to 3.62456 m^3 and $1.28 \times 10^2 \text{ ft}^3$

coulomb

symbol C
system SI (additional unit)
classification unit of electric charge, electric flux, and elementary charge

coulomb meter

symbol $C \cdot m$
system SI
classification unit of electric dipole moment

coulomb meter squared per kilogram

symbol $C \cdot m^2/kg$
system SI
classification unit of specific gamma ray constant

coulomb meter squared per volt

symbol $C \cdot m^2/V$
system SI
classification unit of polarizability of molecule

coulomb per cubic meter

symbol C/m^3
system SI
classification unit of volume density of electric charge

coulomb per kilogram

symbol	C/kg
system	SI
classification	unit of exposure

coulomb per kilogram second

symbol	$C/(kg \cdot s)$
system	SI
classification	unit of rate of exposure

coulomb per mole

symbol	C/mol
system	SI
classification	unit of Faraday constant

coulomb per square meter

symbol	C/m^2
system	SI
classification	unit of surface density of charge, electric flux density, and electric polarization

crocodile (not in use)

classification	unit of electric potential

cubic centimeter

symbol	cm^3
system	SI (multiple unit) and CGS
classification	unit of volume

cubic centimeter per gram

symbol	cm^3/g
system	SI (multiple unit) and CGS
classification	unit of specific volume

cubic centimeter per kilogram

symbol	cm^3/kg
system	SI (multiple unit)
classification	unit of specific volume

cubic decimeter

symbol	dm^3
system	SI (multiple unit)
classification	unit of volume

cubic decimeter per kilogram

symbol	dm^3/kg
system	SI (multiple unit)
classification	unit of specific volume

cubic foot

symbol	ft^3
system	imperial unit
classification	unit of volume
country	United States, United Kingdom

cubic foot per pound

symbol	ft^3/lb
system	foot-pound-second
classification	unit of specific volume

cubic foot per second (cusec)

symbol	ft^3/s
system	foot-pound-second
classification	unit of rate of volume flow

cubic foot per ton

symbol	$ft^3/UKton$
classification	unit of specific volume
country	United Kingdom

cubic inch

symbol	in^3
system	imperial unit
classification	unit of volume
country	United States, United Kingdom

cubic inch per pound

symbol in^3/lb
classification unit of specific volume

cubic meter

symbol m^3
system SI
classification unit of volume

cubic meter per coulomb

symbol m^3/C
system SI
classification unit of Hall coefficient

cubic meter per hour

symbol m^3/h
system non-SI (approved)
classification unit of rate of volume flow

cubic meter per kilogram

symbol m^3/kg
system SI
classification unit of specific volume

cubic meter per mole

symbol m^3/mol
system SI
classification unit of molar volume

cubic meter per second

symbol m^3/s
system SI
classification unit of rate of volume flow

cubic yard

symbol yd^3
system imperial unit

classification	unit of volume
country	United States, United Kingdom

curie

symbol	Ci
classification	unit of activity of radionuclide

curie MeV (not in use)

symbol	Ci · MeV
classification	unit of nuclear power

curie per cubic meter

symbol	Ci/m^3
classification	unit of activity of volume

curie per kilogram

symbol	Ci/kg
classification	unit of specific activity of radionuclide

cycle per second

symbol	c/s
classification	unit of frequency

dalton

classification	unit used for the atomic mass unit

daraf (not in use)

classification	unit equal to 1/farad
country	United States

darcy

symbol	D
classification	unit of permeability

day

symbol	d
system	non-SI (approved)
classification	unit of time

debye (not in use)

symbol D
classification unit of electric dipole moment

decibel

symbol dB
classification unit of sound power level, sound pressure level, sound intensity level, sound reduction index, amplitude level difference, power level difference (dimensionless quantities); equal to 10 bels

decimilligrade (not in use)

symbol $\ldots^{cc}$
classification unit of plane angle equal to centesimal second

degree

symbol $\ldots^{\circ}$
system non-SI (approved)
classification unit of plane angle

degree (not in use)

symbol deg
classification unit of temperature interval

degree absolute (not in use)

classification unit of temperature interval, used for kelvin scale

degree Celsius

symbol °C
system SI (additional unit)
classification unit of temperature interval, used for Celsius scale

degree Centigrade (not in use)

classification unit of temperature interval, used for degree Celsius

degree Fahrenheit

symbol °F
classification unit of temperature interval, used for Fahrenheit scale

degree Kelvin (not in use)

symbol	°K
system	SI
classification	unit of temperature interval, used for kelvin scale, changed to kelvin by Conference Generale des Poids et Mesures in 1967

degree per second

symbol	°/s
classification	unit of angular velocity

degree per second squared

symbol	$°/s^2$
classification	unit of angular acceleration

degree Rankine

symbol	°R
classification	unit of temperature interval, used for Rankine scale

degree Reaumur (not in use)

symbol	°R
classification	unit of temperature interval, used for Reaumur scale

denier (not in use)

symbol	den
classification	unit of linear density

Dezitonne

(*see* quintal)

dioptre

symbol	δ, dpt
classification	unit of lenses (optics)

drachm, apothecaries'

system	apothecaries' unit
classification	unit of fluid volume equal to 4.61395 in^3
country	United Kingdom

dram, apothecaries'

symbol	dr ap
system	apothecaries' unit
classification	unit of mass equal to 3.88793×10^{-3} kg
country	United States

dram, avoirdupois

symbol	dr
system	avoirdupois unit
classification	unit of mass equal to 1.77185×10^{-3} kg
country	United States, United Kingdom

drex

classification	unit of density
country	Canada, United States

dry barrel

symbol	bbl
classification	unit of volume, used for dry goods, equal to 1.15627×10^{-1} m^3
country	United States

dry pint

symbol	dry pt
classification	unit of volume, used for dry goods, equal to 5.50610×10^{-4} m^3
country	United States

dry quart

symbol	dry qt
classification	unit of volume, used for dry goods, equal to 1.10122 dm^3
country	United States

dyne

symbol	dyn
system	CGS
classification	unit of force equal to one gram × centimeter/square second

dyne centimeter

symbol dyn · cm
system CGS
classification unit of moment of force

dyne centimeter per biot

symbol dyn · cm/Bi
system CGS
classification unit of magnetic flux

dyne centimeter per second

symbol dyn · cm/s
system CGS
classification unit of moment of momentum

dyne per biot centimeter

symbol dyn/(Bi · cm)
system CGS
classification unit of magnetic flux density and magnetic polarization

dyne per biot squared

symbol dyn/Bi^2
system CGS
classification unit of permeability

dyne per centimeter

symbol dyn/cm
system CGS
classification unit of surface tension

dyne per cubic centimeter

symbol dyn/cm^3
system CGS
classification unit of specific weight

dyne per franklin

symbol dyn/Fr
system CGS
classification unit of strength of electric field

dyne per square centimeter

symbol dyn/cm^2
system CGS
classification unit of pressure

dyne second

symbol dyn · s
system CGS
classification unit of momentum

dyne second per centimeter

symbol dyn · s/cm
system CGS
classification unit of mechanical impedance

dyne second per centimeter cubed

symbol dyn · s/cm^3
system CGS
classification unit of specific impedance in acoustics

dyne second per centimeter to the fifth power

symbol dyn · s/cm^5
system CGS
classification unit of impedance in acoustics

dyne second per square centimeter

symbol dyn · s/cm^2
(*see* poise)

electronvolt

symbol eV
system non-SI (approved)
classification unit of energy

electronvolt per meter

symbol eV/m
system non-SI (approved)
classification unit of linear stopping power and linear energy transfer

electronvolt per square meter

symbol	eV/m^2
system	non-SI (approved)
classification	unit of energy fluence

electronvolt per square meter second

symbol	$eV/(m^2 \cdot s)$
system	non-SI (approved)
classification	unit of rate of energy fluence

electronvolt square meter

symbol	$eV \cdot m^2$
system	non-SI (approved)
classification	unit of atomic stopping power

electronvolt square meter per kilogram

symbol	$eV \cdot m^2/kg$
system	non-SI (approved)
classification	unit of mass stopping power

engineer's chain

classification	unit of length equal to 3.048×10 m and 1.0×10^2 ft

erg

symbol	erg
system	CGS
classification	unit of work and energy

erg per biot

symbol	erg/Bi
system	CGS
classification	unit of magnetic flux

erg per biot squared

symbol	erg/Bi^2
system	CGS
classification	unit of self inductance and mutual inductance

erg per centimeter

symbol erg/cm
system CGS
classification unit of linear stopping power and linear energy transfer

erg per cubic centimeter

symbol erg/cm^3
system CGS
classification unit of energy density and calorific value per unit of volume

erg per cubic centimeter degree Celsius

symbol $erg/(cm^3 \cdot °C)$
system CGS
classification unit of heat capacity per unit volume

erg per cubic centimeter second

symbol $erg/(cm^3 \cdot s)$
system CGS
classification unit of rate of heat flow

erg per centimeter second degree Celsius

symbol $erg/(cm \cdot s \cdot °C)$
system CGS
classification unit of thermal conductivity

erg per degree Celsius

symbol erg/°C
(*see* erg per kelvin)

erg per franklin

symbol erg/Fr
system CGS
classification unit of electric potential

erg per gram

symbol erg/g
system CGS

classification	unit of energy per unit weight, or specific energy, or kinetic energy per unit weight (kerma, gray, or absorbed dose)

erg per gram degree Celsius

symbol	erg/(g · °C)
system	CGS
classification	unit of specific heat capacity

erg per gram second

symbol	erg/(g · s)
system	CGS
classification	unit of rate of absorbed dose, or rate of kerma, or rate of gray, or rate of energy per unit weight

erg per kelvin

symbol	erg/K
system	CGS
classification	unit of heat capacity and entropy

erg per mole degree Celsius

symbol	erg/(mol · °C)
system	CGS
classification	unit of molar gas constant

erg per second

symbol	erg/s
system	CGS
classification	unit of power and sound energy flux

erg per second steradian

symbol	erg/(s · sr)
system	CGS
classification	unit of radiant intensity

erg per second steradian square centimeter

symbol	erg/(s · sr · cm^2)
system	CGS
classification	unit of radiance

erg per square centimeter

symbol erg/cm^2
(*see* dyne per centimeter)

erg per square centimeter second

symbol $erg/(cm^2 \cdot s)$
system CGS
classification unit of rate of energy fluence

erg per square centimeter second degree Celsius

symbol $erg/(cm^2 \cdot s \cdot °C)$
system CGS
classification unit of coefficient of heat transfer

erg per square centimeter second kelvin to the fourth power

symbol $erg/(cm^2 \cdot s \cdot K^4)$
system CGS
classification unit of Stefan-Boltzmann constant

erg second

symbol $erg \cdot s$
system CGS
classification unit of Planck constant

erg square centimeter

symbol $erg \cdot cm^2$
system CGS
classification unit of atomic stopping power

erg square centimeter per gram

symbol $erg \cdot cm^2/g$
system CGS
classification unit of first radiation constant

farad

symbol F
system SI (additional unit)
classification unit of capacitance

farad per meter

symbol F/m
system SI
classification unit of permittivity

farad square meter

symbol $F \cdot m^2$
(*see* coulomb meter squared per volt)

fathom

classification unit of length (nautical)

fermi

classification unit of length in nuclear physics

fluid drachm

symbol UK fl dr
classification unit of volume equal to 3.55163×10^{-6} m^3
country United Kingdom

fluid dram (not in use)

symbol US fl dr
classification unit of volume equal to 3.69669×10^{-6} m^3, used for measuring liquids
country United States

fluid ounce

symbol UK fl oz
system imperial unit
classification unit of volume equal to 2.84131×10^{-5} m^3
country United Kingdom

fluid ounce (*liquid ounce*)

classification unit of volume equal to 2.95735×10^{-5} m^3, used for measuring liquids
country United States

foot

symbol ft
system foot-pound-second and imperial unit
classification unit of length
country United States, United Kingdom

foot, board

classification unit of volume used for wood products equal to 12 in × 12 in × 1 in
(*see* board foot)

foot cubed

symbol ft^3
classification unit of section modulus (structural engineering)
(*see* cubic foot)

foot hour degree Fahrenheit per British thermal unit

symbol ft · h · °F/Btu
classification unit of thermal resistivity

foot of water (conventional)

symbol ftH_2O
classification unit of pressure

foot per minute

symbol ft/min
classification unit of velocity

foot per pound

symbol ft/lb
system foot-pound-second
classification unit of specific length

foot per second

symbol ft/s
system foot-pound-second
classification unit of velocity

foot per second squared

symbol ft/s^2
system foot-pound-second
classification unit of acceleration

foot poundal

symbol ft · pdl
(*see* poundal foot)

foot poundal per second

system foot-pound-second
classification unit of power

foot pound-force

symbol ft · lbf
system foot-pound force-second
classification unit of work

foot pound-force per pound

symbol ft · lbf/lb
system foot-pound force-second
classification unit of specific internal energy and specific latent heat

foot pound-force per pound degree Fahrenheit

symbol ft · lbf/(lb · °F)
system foot-pound force-second
classification unit of specific heat capacity

foot pound-force per second

symbol ft · lbf/s
system foot-pound force-second
classification unit of power

foot squared per hour

symbol ft^2/h
classification unit of kinematic viscosity

foot squared per second

symbol	ft^2/s
system	foot-pound-second
classification	unit of kinematic viscosity

foot to the fourth power

symbol	ft^4
system	foot-pound-second
classification	unit of second moment of area

foot-candle

symbol	fc
classification	unit of illuminance

(*see* lumen per square foot)

foot-lambert

symbol	ft · La
classification	unit of luminance

franklin

symbol	Fr
system	CGS
classification	unit of electric charge and electric flux

franklin centimeter

symbol	Fr · cm
system	CGS
classification	unit of electric dipole moment

franklin per second

symbol	Fr/s
system	CGS
classification	unit of electric current

franklin per square centimeter

symbol	Fr/cm^2
system	CGS
classification	unit of polarization and electric flux density

franklin squared per erg

symbol Fr^2/erg
system CGS
classification unit of capacitance

franklin squared per erg centimeter

symbol $Fr^2/(erg \cdot cm)$
system CGS
classification unit of permittivity

freight ton

classification unit used in shipping equal to 40 ft^3 and 1.132674 m^3

frigorie

symbol fg
classification unit of heat for refrigeration equal to 1.0 kcal

frigorie per hour

symbol fg/h
classification unit of refrigerating capacity

furlong

system imperial unit
classification unit of length equal to 2.01168×10^2 m
country United States, United Kingdom

gal

symbol Gal
system CGS
classification unit of acceleration (linear) equal to 1 cm/s^2

*gallon**

symbol UKgal
system imperial unit
classification unit of volume equal to 4.54609 dm^3 or L
country United Kingdom

**Series of U.K. gallons precedes series of U.S. gallons here.*

gallon per hour

symbol UKgal/h
classification unit of rate of volume flow equal to 4.54609×10^{-3} m^3/h
country United Kingdom

gallon per mile

symbol UKgal/mile
classification unit of fuel consumption equal to 2.82481 liters/km
country United Kingdom

gallon per minute

symbol UKgal/min
classification unit of rate of volume flow equal to 7.57682×10^{-5} m^3/s
country United Kingdom

gallon per pound

symbol UKgal/lb
classification unit of specific volume equal to 1.00224×10^{-2} m^3/kg
country United Kingdom

gallon per second

symbol UKgal/s
classification unit of volume equal to 4.54609×10^{-3} m^3/s
country United Kingdom

gallon

symbol USgal
classification unit of volume equal to 3.78541 dm^3 or L
country United States

gallon per hour

symbol USgal/h
classification unit of rate of volume flow equal to 3.78541×10^{-3} m^3/h
country United States

gallon per mile

symbol USgal/mile
classification unit of fuel consumption equal to 2.35215 liters/km
country United States

gallon per minute

symbol USgal/min
classification unit of rate of volume flow equal to 6.30902×10^{-5} m^3/s
country United States

gallon per pound

symbol USgal/lb
classification unit of specific volume equal to 8.34540×10^{-3} m^3/kg
country United States

gallon per second

symbol USgal/s
classification unit of rate of volume flow equal to 3.78541×10^{-3} m^3/s
country United States

gamma

symbol γ
classification unit of mass and unit of magnetic flux density

gauss

symbol Gs, G
system electromagnetic CGS
classification unit of density of magnetic flux

gee pound

(*see* slug)

gilbert

symbol Gb
system electromagnetic CGS
classification unit of magnetomotive force

gilbert per centimeter

symbol Gb/cm
system electromagnetic CGS
classification unit of strength of magnetic field

gilbert per maxwell

symbol Gb/Mx
system electromagnetic CGS
classification unit of reluctance, equal to 1/henry and equal to 1/permeance

gill

system imperial unit
classification unit of volume equal to 1.42065×10^{-4} m^3
country United Kingdom

gill

symbol gi
classification unit of volume equal to 1.18294×10^{-4} m^3, used for measuring liquids
country United States

gon

symbol $\ldots^{g}$
classification unit of plane angle obtained by dividing 90° into one hundred parts

grade

symbol $\ldots^{g}$
classification unit of plane angle. One grade equals one gon; 100 grades equals 90°; 400 grades equals 360°.

grade per second

symbol g/s
classification unit of angular velocity

grade per second squared

symbol g/s^2
classification unit of angular acceleration

grain

symbol gr
system apothecaries' unit, avoirdupois unit, imperial unit, and troy unit
classification unit of mass equal to 6.479891 × 10 milligrams
country United States, United Kingdom

grain per cubic foot

symbol gr/ft^3
classification unit of density and concentration

grain per gallon

symbol gr/UKgal
classification unit of density and concentration equal to 1.42538×10^{-2} kg/m^3
country United Kingdom

grain per gallon

symbol gr/UKgal
classification unit of density and concentration equal to 1.71181×10^{-2} kg/m^3
country United States

gram

symbol g
system SI (multiple unit) and CGS (base unit)
classification unit of mass

gram centimeter per second

symbol g · cm/s
system CGS
classification unit of momentum

gram centimeter per second squared

symbol $g \cdot cm/s^2$
(*see* dyne)

gram centimeter squared

symbol $g \cdot cm^2$
system CGS
classification unit of moment of inertia

gram centimeter squared per second

symbol $g \cdot cm^2/s$
system CGS
classification unit of moment of momentum

gram per cubic centimeter

symbol g/cm^3
system CGS
classification unit of density (mass)

gram per liter

symbol g/l
classification unit of density (mass)

gram per milliliter

symbol g/ml
classification unit of density (mass)

gram per square meter

symbol g/m^2
system SI (multiple unit)
classification unit of density (surface)

gram per square meter day

symbol $g/(m^2 \cdot d)$
system non-SI (approved)
classification unit of rate of transfer of water vapor

gram-atom (not in use)

classification unit of mass of an element

gram-calorie (not in use)

classification name for calorie

gram-force

symbol gf
system meter-kilogram force-second
classification unit of force

gram-molecule (not in use)

symbol gmol
classification unit of mass of a compound

gram-rad

symbol g · rad
classification unit of integral absorbed dose (where absorbed dose equals gray, equals kerma, and equals kinetic energy per unit weight)

gram-weight (not in use)

symbol g, g(wt)
classification unit same as gram-force

gray

symbol Gy
system SI (additional unit)
classification unit of energy per unit weight, specific energy, or kinetic energy per unit weight (kerma or absorbed dose)

gray per second

symbol Gy/s
system SI
classification unit of rate of absorbed dose, rate of kerma, or rate of energy per unit weight

hand

classification unit of length (old system used to measure the height of horses)

hectare

symbol ha
classification unit of area equal to 10^4 m^2, 10^2 are, 10^{-2} km^2, used in land surveying

hectare-millimeter

symbol ha · mm
classification unit of volume equal to 10 m^3

hectoliter

symbol hl
classification unit of volume equal to 10^2 dm^3, used in brewing manufacturing

hectopièze

symbol hpz
classification French unit of pressure equal to one bar

Hefner candle (not in use)

classification unit of luminous intensity equal to 0.903 candela (before 1942 used prominently in Germany)

henry

symbol H
system SI (additional unit)
classification unit of permeance, self inductance, and mutual inductance

henry per meter

symbol H/m
system SI
classification unit of permeability

hertz

symbol	Hz
system	SI (additional unit)
classification	unit of frequency

horsepower

symbol	hp
classification	unit of power equal to 1.01387 metric horsepower
country	United States, United Kingdom

horsepower (metric)

classification	unit of power equal to 9.86320×10^{-1} horsepower

horsepower hour

symbol	$hp \cdot h$
classification	unit of energy equal to 7.45700×10^{-1} $kW \cdot h$
country	United States, United Kingdom

horsepower hour (metric)

classification	unit of work equal to 7.35499×10^{-1} $kW \cdot h$

hour

symbol	h
system	non-SI (approved)
classification	unit of time

hundredweight

symbol	cwt
system	avoirdupois unit and imperial unit
classification	unit of mass equal to 5.08023×10 kg
country	United Kingdom

hyl

system	meter-kilogram force-second
classification	unit of mass equal to 9.80665×10^{-3} kg

inch

symbol in
system imperial unit
classification unit of length equal to 25.4 millimeters
country United States, United Kingdom

inch cubed

symbol in^3
classification unit of section modulus

inch of mercury

symbol inHg
classification unit of pressure

inch of water

symbol inH_2O
classification unit of pressure

inch per minute

symbol in/min
classification unit of velocity

inch per second

symbol in/s
classification unit of velocity

inch squared per hour

symbol in^2/h
classification unit of kinematic viscosity

inch squared per second

symbol in^2/s
classification unit of kinematic viscosity

inch to the fourth power

symbol in^4
classification unit of second moment of area

inhour

classification — unit of reactivity equal to the increase in reactivity of a critical reactor that produces a reactor time constant of one hour (dimensionless quantity)

international ampere (not in use)

symbol — A_{int}
classification — unit of electric current equal to 9.9985×10^{-1} ampere

international candle (not in use)

symbol — IC
classification — unit of luminous intensity equal to 1.02 candelas

international coulomb (not in use)

symbol — C_{int}
classification — unit of electric charge equal to 9.9985×10^{-1} coulomb

international farad (not in use)

symbol — F_{int}
classification — unit of capacitance equal to 9.9951×10^{-1} farad

international henry (not in use)

symbol — H_{int}
classification — unit of inductance and permeance equal to 1.00049 henrys

international joule (mean) (not in use)

symbol — J_{int}
classification — unit of work, energy, and heat equal to 1.00019 joules

international ohm (not in use)

symbol — Ω_{int}
classification — unit of resistance equal to 1.00049 Ω

international siemens (not in use)

symbol — S_{int}
classification — unit of conductance equal to 9.9951×10^{-1} siemens

international table calorie

symbol cal_{IT}
classification unit of heat

international table kilocalorie

symbol $kcal_{IT}$
classification unit of heat

international tesla (not in use)

symbol T_{int}
classification unit of magnetic flux density equal to 1.00034 teslas

international volt (not in use)

symbol V_{int}
classification unit of electric potential

international watt (not in use)

symbol W_{int}
classification unit of power equal to 1.00019 watts

international weber (not in use)

symbol Wb_{int}
classification unit of magnetic flux equal to 1.00034 webers

joule

symbol J
system SI (additional unit)
classification unit of work and energy

joule per cubic meter

symbol J/m^3
system SI
classification unit of energy density

joule per kelvin

symbol J/K
system SI
classification unit of heat capacity and entropy

joule per kilogram

symbol J/kg
system SI
classification unit of specific energy and specific enthalpy

joule per kilogram kelvin

symbol J/(kg · K)
system SI
classification unit of specific heat capacity and specific entropy

joule per kilogram second

symbol J/(kg · s)
(*see* watt per kilogram)

joule per meter

symbol J/m
system SI
classification unit of linear stopping power and linear energy transfer

joule per meter to the fourth power

symbol J/m
system SI
classification unit of spectral concentration of density of radiant energy

joule per mole

symbol J/mol
system SI
classification unit of molar internal energy

joule per mole kelvin

symbol J/(mol · K)
system SI
classification unit of molar heat capacity and molar entropy

joule per pound kelvin or degree Celsius

symbol J/(lb · K), J/(lb · °C)
classification unit of specific heat capacity and specific entropy

joule per second

symbol J/s
(*see* watt)

joule per square meter

symbol J/m^2
system SI
classification unit of energy fluence and radiant exposure

joule per square meter second

symbol $J/(m^2 \cdot s)$
(*see* watt per square meter)

joule per square meter second kelvin

symbol $J/(m^2 \cdot s \cdot K)$
(*see* watt per square meter kelvin)

joule per tesla

symbol J/T
(*see* ampere square meter)

joule reciprocal hertz

symbol $J \cdot Hz^{-1}$
(*see* joule second)

joule reciprocal tesla

symbol $J \cdot T^{-1}$
(*see* ampere square meter)

joule second

symbol $J \cdot s$
system SI
classification unit of Planck constant and action

joule square meter

symbol $J \cdot m^2$
system SI
classification unit of atomic stopping power

joule square meter per kilogram

symbol $J \cdot m^2/kg$
system SI
classification unit of mass stopping power

Julian year

classification unit of time equal to 3.6525×10^2 days

kayser (not in use)

symbol K
classification unit of wave number

kelvin

symbol K
system SI (base unit)
classification unit of thermodynamic temperature and unit of temperature interval and other temperatures

kelvin per meter

symbol K/m
system SI
classification unit of temperature gradient

kelvin per watt

symbol K/W
system SI
classification unit of thermal resistance

kilocalorie (I.T.) (not in use)

symbol $kcal_{IT}$ or kcal
classification unit of heat

kilocalorie (I.T.) meter per square meter hour kelvin or degree Celsius

symbol $kcal_{IT} \cdot m/(m^2 \cdot h \cdot K)$ or $kcal_{IT} \cdot m/(m^2 \cdot h \cdot °C)$
classification unit of thermal conductivity

kilocalorie (I.T.) per cubic meter

symbol $kcal_{IT}/m^3$
classification unit of calorific value per unit of volume

kilocalorie (I.T.) per cubic meter hour

symbol $kcal_{IT}/(m^3 \cdot h)$
classification unit of rate of heat flow

kilocalorie (I.T.) per hour

symbol $kcal_{IT}/h$
classification unit of rate of heat flow

kilocalorie (I.T.) per kelvin or degree Celsius

symbol $kcal_{IT}/K$, $kcal_{IT}/{}^\circ C$
classification unit of heat capacity

kilocalorie (I.T.) per kilogram

symbol $kcal_{IT}/kg$
classification unit of specific internal energy and calorific value per unit of mass

kilocalorie (I.T.) per kilogram kelvin or degree Celsius

symbol $kcal_{IT}/(kg \cdot K)$ or $kcal_{IT}/(kg \cdot {}^\circ C)$
classification unit of heat capacity

kilocalorie (I.T.) per meter hour kelvin or degree Celsius

symbol $kcal_{IT}/(m \cdot h \cdot K)$, $kcal_{IT}/(m \cdot h \cdot {}^\circ C)$
classification unit of thermal conductivity

kilocalorie (I.T.) per square meter hour

symbol $kcal_{IT}/(m^2 \cdot h)$
classification unit of density of rate of heat flow

kilocalorie (I.T.) per square meter hour kelvin or degree Celsius

symbol $kcal_{IT}/(m^2 \cdot h \cdot K)$, $kcal_{IT}/(m^2 \cdot h \cdot {}^\circ C)$
classification unit of coefficient of heat transfer

kilogram

symbol	kg
system	SI (base unit)
classification	unit of mass

kilogram meter per second

symbol	$kg \cdot m/s$
system	SI
classification	unit of momentum

kilogram meter per second squared

symbol $kg \cdot m/s^2$
(*see* newton)

kilogram meter squared

symbol	$kg \cdot m^2$
system	SI
classification	unit of moment of inertia

kilogram meter squared per second

symbol	$kg \cdot m^2/s$
system	SI
classification	unit of moment of momentum

kilogram per cubic centimeter

symbol	kg/cm^3
system	SI (multiple unit)
classification	unit of density (mass)

kilogram per cubic decimeter

symbol	kg/dm^3
system	SI (multiple unit)
classification	unit of density (mass)

kilogram per cubic meter

symbol	kg/m^3
system	SI
classification	unit of density (mass)

kilogram per cubic meter pascal

symbol kg/(m^3 · Pa)
system SI
classification unit of unitary mass density

kilogram per hectare

symbol kg/ha
classification unit of density (surface)

kilogram per hour

symbol kg/h
system non-SI (approved)
classification unit of rate of mass flow

kilogram per liter

symbol kg/l, kg/L
system non-SI (approved)
classification unit of density (mass)

kilogram per meter

symbol kg/m
system SI
classification unit of density (linear)

kilogram per meter second

symbol kg/(m · s)
(*see* pascal second)

kilogram per mole

symbol kg/mol
system SI
classification unit of molar mass

kilogram per pascal second meter

symbol kg/(Pa · s · m)
system SI
classification unit of water vapor permeance

kilogram per pascal second square meter

symbol — $kg/(Pa \cdot s \cdot m^2)$
system — SI
classification — unit of water vapor permeability

kilogram per second

symbol — kg/s
system — SI
classification — unit of rate of mass flow

kilogram per square centimeter

symbol — kg/cm^2
system — SI (multiple unit)
classification — unit of density (surface)

kilogram per square meter

symbol — kg/m^2
system — SI
classification — unit of density (surface)

kilogram-calorie (not in use)

symbol — kcal
classification — unit same as kilocalorie

kilogram-force

symbol — kgf
system — meter-kilogram force-second (base unit)
classification — unit of force ($kgf = kg\ (mass) \times g$; $kgf = N \cdot g$)

kilogram-force meter

symbol — $kgf \cdot m$
system — meter-kilogram force-second
classification — unit of moment of force and torque; work and energy

kilogram-force meter per kilogram

symbol — $kgf \cdot m/kg$
system — meter-kilogram force-second
classification — unit of specific internal energy and specific latent heat

kilogram-force meter per kilogram degree Celsius

symbol	kgf · m/kg · °C
system	meter-kilogram force-second
classification	unit of specific heat capacity

kilogram-force meter per second

symbol	kgf · m/s
system	meter-kilogram force-second
classification	unit of power

kilogram-force meter second

symbol	kgf · m · s
system	meter-kilogram force-second
classification	unit of action

kilogram-force meter second squared

symbol	kgf · m · s^2
system	meter-kilogram force-second
classification	unit of moment of inertia

kilogram-force per centimeter

symbol	kgf/cm
system	meter-kilogram force-second
classification	unit of surface tension

kilogram-force per cubic meter

symbol	kgf/m^3
system	meter-kilogram force-second
classification	unit of specific weight

kilogram-force per meter

symbol	kgf/m
system	meter-kilogram force-second
classification	unit of surface tension

kilogram-force per meter second degree Celsius

symbol	kgf/(m·s·°C)
system	meter-kilogram force-second
classification	unit of coefficient of heat transfer

kilogram-force per second degree Celsius

symbol kgf/s·°C
system meter-kilogram force-second
classification unit of thermal conductivity

kilogram-force per square centimeter

symbol kgf/cm^2
system meter-kilogram force-second
classification unit of pressure

kilogram-force per square meter

symbol kgf/m^2
system meter-kilogram force-second
classification unit of pressure

kilogram-force second

symbol kgf · s
system meter-kilogram force-second
classification unit of momentum

kilogram-force second per square meter

symbol $kgf \cdot s/m^2$
system meter-kilogram force-second
classification unit of dynamic viscosity

kilogram-force second squared per meter

symbol $kgf \cdot s^2/m$
system meter-kilogram force-second
classification unit of mass equal to 9.80665 kg

kilogram-force second squared per meter to the fourth power

symbol $kgf \cdot s^2/m^4$
system meter-kilogram force-second
classification unit of density

kilogram-weight (not in use)

symbol kg, kg (wt)
classification unit same as kilogram-force

kilohl

symbol	khyl
system	meter-kilogram force-second
classification	unit of mass

kilohyl per cubic meter

symbol	$khyl/m^3$
classification	unit of density

kilometer

symbol	km
system	SI (multiple unit)
classification	unit of length

kilometer per hour

symbol	km/h
system	non-SI (approved)
classification	unit of velocity

kilomole

symbol	kmol
classification	unit equal to 10^3 moles

kilopond (not in use)

symbol	kp
system	meter-kilopond-second
classification	unit of force synonymous with kilogram-force (used in Central Europe)

kilowatt

symbol	kW
system	SI (multiple unit)
classification	unit of power

kilowatt hour

symbol	$kW \cdot h$
system	non-SI (approved)
classification	unit of energy

kip

classification unit of force equal to 10^3 pound-force
country United States

knot

classification unit of velocity equal to 1.85318 km/h and 1.00064 international knots
country United Kingdom

knot (international)

symbol kn
classification unit of velocity equal to 1.852 km/h and 9.99361×10^{-1} UKknot

lambda (not in use)

symbol λ
classification unit of volume

lambert

symbol La
classification unit of luminance equal to 3.18310×10^3 candelas/square meter

langley

classification unit of surface density of radiant energy equal to one calorie (I.T.)/square centimeter

langley per minute

symbol langley/min
classification unit of irradiance

light year

symbol l.y.
classification unit of length equal to 9.4607×10^{15} m

link

classification unit of length equal to 2.01168×10^{-1} m and 7.92 in

liquid ounce

symbol USliq oz
classification unit of volume equal to 2.95735×10^{-2} dm^3 or liter and 1.80469 in^3, used for measuring liquids
country United States

liquid pint

symbol USliq pt
classification unit of volume equal to 4.73176×10^{-1} dm^3 or liter and 2.8875×10 in^3, used for measuring liquids
country United States

liquid quart

symbol USliq qt
classification unit of volume equal to 9.46353×10^{-4} m^3, used for measuring liquids
country United States

liter

symbol l, L
system non-SI (approved)
classification unit of volume equal to 10^{-3} m^3

liter (old) (not in use)

symbol l
classification unit of volume equal to 1.000028×10^{-3} m^3, used between 1901 and 1964

liter atmosphere

symbol $l \cdot atm$
classification unit of work

liter per 100 kilometers

symbol l/100 km, L/100 km
system non-SI (approved)
classification unit of fuel consumption

liter per kilogram

symbol l/kg, L/kg
system non-SI (approved)
classification unit of specific volume

liter per mole

symbol l/mol, L/mol
system non-SI (approved)
classification unit of molar volume

liter per second

symbol l/s, L/s
system non-SI (approved)
classification unit of rate of volume flow

lumen

symbol lm
system SI (additional unit)
classification unit of luminous flux

lumen hour

symbol lm · h
system non-SI (approved)
classification unit of quantity of light

lumen per square foot

symbol lm/ft^2
classification unit of illuminance

lumen per square meter

symbol lm/m^2
system SI
classification unit of luminous exitance

lumen per watt

symbol lm/W
system SI
classification unit of luminous efficacy

lumen second

symbol	lm · s
system	SI
classification	unit of quantity of light

lusec

classification	unit of fluid escape rate, used in association with vacuum measurements

lux

symbol	lx
system	SI (additional unit)
classification	unit of illuminance

lux hour

symbol	lx · h
system	non-SI (approved)
classification	unit of light exposure

lux second

symbol	lx·s
system	SI
classification	unit of light exposure

Mach number

symbol	Ma, M
classification	Mach number is the ratio of the velocity of an object or fluid to the velocity of sound in the same medium and under the same conditions.

magnetic ohm

classification	unit used for gilbert per maxwell

maxwell

symbol	Mx, M
system	electromagnetic CGS
classification	unit that measures the magnetic flux that produces an electromotive force of one abvolt in a circuit of one turn

linking the flux, as the flux is reduced to zero in one second at a uniform rate, or measures the amount of flux passing through one square centimeter normal to a magnetic field with an intensity of one gauss

maxwell per square centimeter

symbol Mx/cm^2
(*see* gauss)

mechanical ohm (not in use)

classification unit of mechanical impedance

megagram

symbol Mg
system SI (multiple unit)
classification unit of mass equal to 10^3 kg

megapascal

symbol MPa
system SI (multiple unit)
classification unit of pressure or stress equal to 10^6 pascals

megapond

symbol Mp
classification unit of force equal to 10^3 kiloponds

meter

symbol m
system SI (base unit)
classification unit of length

meter cubed

symbol m^3
system SI
classification unit of section modulus

meter hour degree Celsius per kilocalorie (I.T.)

symbol $m \cdot h \cdot °C/kcal_{IT}$
classification unit of thermal resistivity

meter kelvin

symbol	m · K
system	SI
classification	unit of second radiation constant

meter kelvin per watt

symbol	m · K/W
system	SI
classification	unit of thermal resistivity

meter of water

symbol	mH_2O
classification	unit of pressure equal to 9.80665×10^3 pascals

meter per kilogram

symbol	m/kg
system	SI
classification	unit of specific length

meter per second

symbol	m/s
system	SI
classification	unit of velocity

meter per second cubed

symbol	m/s^3
system	SI
classification	unit of jerk

meter per second squared

symbol	m/s^2
system	SI
classification	unit of acceleration

meter second per kilogram

symbol	m · s/kg

(*see* reciprocal pascal reciprocal second)

meter squared

symbol m^2
system SI
classification unit of migration area, decreasing area, and diffusion area

meter squared per hour

symbol m^2/h
system non-SI (approved)
classification unit of kinematic viscosity

meter squared per newton second

symbol $m^2/(N \cdot s)$
(*see* reciprocal pascal reciprocal second)

meter squared per second

symbol m^2/s
system SI
classification unit of kinematic viscosity

meter to the fourth power

symbol m^4
system SI
classification unit of second moment of area

metric carat

classification unit of mass adopted in 1907 by Conference Generale des Poids et Mesures

metric technical unit of mass

system meter-kilogram force-second
classification unit of mass equal to 9.80665 kg

microbar

symbol μbar
classification unit equal to 10^{-1} pascal

micro-inch

symbol	μin
classification	unit of length equal to 1.0×10^{-6} in

microkatal

symbol	μkat
classification	unit of enzyme activity

micrometer

symbol	μm
system	SI (multiple unit)
classification	unit of length equal to 10^{-6} m

micron (not in use)

symbol	μ
classification	same as micrometer

micron of mercury

symbol	μHg
classification	unit of fluid pressure

microtorr (not in use)

symbol	μTorr
classification	unit of fluid pressure

mile

symbol	mile
system	imperial unit
classification	unit of length
country	United States, United Kingdom

mile per gallon

symbol	mile/UKgal
classification	unit same as 1/fuel consumption and equal to 8.32674 $\times 10^{-1}$ mile/USgal
country	United Kingdom

mile per gallon

symbol mile/USgal
classification unit same as 1/fuel consumption and equal to 1.20095 miles/UKgal
country United States

mile per hour

symbol mile/h
classification unit of velocity
country United States, United Kingdom

millibar

symbol mbar, mb
classification unit of pressure equal to 10^{-3} bar, used in meteorological barometry

milligal

symbol mGal
classification unit of acceleration equal to 10^{-3} gal

milligrade

symbol $...^{mg}$
classification unit of plane angle equal to 10^{-3} grade

milligram per liter

symbol mg/l, mg/L
system non-SI (approved)
classification unit of density and concentration (mass)

milli-inch or mil

symbol min
classification unit of length equal to 1.0×10^{-3} in

milliliter or mil

symbol ml, mL
system non-SI (approved)
classification unit of volume equal to 10^{-3} liter

millimeter

symbol mm
symbol SI (multiple unit)
classification unit of length equal to 10^{-3} m, primarily used in engineering

millimeter of mercury

symbol mmHg
classification unit of pressure equal to 1.33322×10^2 pascals

millimeter of water

symbol mmH_2O
classification unit of pressure equal to 9.80665 pascals

millimicron (not in use)

symbol $m\mu$
classification unit same as nanometer

millitorr (not in use)

symbol mTorr
classification unit of fluid pressure

minim

symbol Ukmin
classification unit of volume equal to 3.61223×10^{-3} in^3
country United Kingdom

minim

symbol USmin
classification unit of volume equal to 3.75977×10^{-3} in^3, used for measuring liquids
country United States

minute

symbol min
system non-SI (approved)
classification unit of time

minute

system non-SI (approved)
classification unit of plane angle

mole

symbol mol
system SI (base unit)
classification unit of amount of substance

mole per cubic meter

symbol mol/m^3
system SI
classification unit of concentration

mole per kilogram

symbol mol/kg
system SI
classification unit of molality and ionic strength

mole per liter

symbol mol/l, mol/L
system non-SI (approved)
classification unit of concentration

mole per second

symbol mol/s
system SI
classification unit of rate of molar flow

nanogram per pascal second square meter

symbol $ng/(Pa \cdot s \cdot m^2)$
system SI (multiple unit)
classification unit used in calculations of moisture transfer in building structures

nanometer

symbol nm
system SI (multiple unit)
classification unit of length equal to 10^{-9} m or one millimicron

nautical mile

classification — unit of length equal to 1.00064 nautical miles (international)
country — United Kingdom

nautical mile (international)

symbol — n mile
classification — unit of length equal to 9.99361×10^{-1} nautical mile, used in United Kingdom

neper

symbol — Np
classification — units of logarithmic decrement, amplitude level difference, and power level difference (dimensionless quantities)

neper per second

symbol — Np/s
classification — unit of damping coefficient equal to $1\ s^{-1}$

newton

symbol — N
system — SI (additional unit)
classification — unit of force ($N = kg$ (mass) $\times 1\ m/sec^2$; $N = kg$ (force)/g)

newton meter

symbol — $N \cdot m$
system — SI
classification — unit of moment of force and torque

newton meter per second

symbol — $N \cdot m/s$
(*see* pascal cubic meter per second)

newton meter second

symbol — $N \cdot m \cdot s$
(*see* kilogram meter squared per second)

newton meter squared per ampere

symbol $N \cdot m^2/A$
(*see* weber meter)

newton per cubic meter

symbol N/m^3
classification unit of specific weight

newton per meter

symbol N/m
system SI
classification unit of surface tension

newton per meter cubed

symbol N/m^3
(*see* pascal per meter)

newton per square meter

symbol N/m^2
(*see* pascal)

newton per weber

symbol N/Wb
(*see* ampere per meter)

newton second

symbol $N \cdot s$
(*see* kilogram meter per second)

newton second per meter

symbol $N \cdot s/m$
system SI
classification unit of mechanical impedance

newton second per meter cubed

symbol $N \cdot s/m^3$
(*see* pascal second per meter)

newton second per meter squared

symbol N · s/m^2
(*see* pascal second)

newton second per square meter

symbol N · s/m^2
(*see* newton second per meter squared)

newton second to the fifth power

symbol N · s/m^5
(*see* pascal second per meter cubed)

newton square meter per ampere

symbol N · m^2/A
(*see* weber meter)

newton square meter per kilogram squared

symbol N · m^2/kg^2
system SI
classification unit of gravitational constant

nile

classification unit of reactivity (dimensionless quantity)

nit

symbol nt
classification unit of luminance equal to one candela per square meter

normal atmosphere

(*see* atmosphere (standard))

octant

classification unit of plane angle

octave

classification unit of frequency interval (dimensionless quantity) *octet*

(*see* byte)

oersted

symbol	Oe
system	electromagnetic CGS
classification	unit of strength of magnetic field

ohm

symbol	Ω
system	SI (additional unit)
classification	unit of impedance, impedance modulus, reactance, and resistance

ohm circular mil per foot

symbol	Ω · circ · mil/ft
classification	unit of resistivity

ohm meter

symbol	Ω · m
system	SI
classification	unit of resistivity

ohm second

symbol	Ω · s

(*see* henry)

ohm square millimeter per meter

symbol	Ω · mm^2/m
classification	unit of resistivity

ounce, apothecaries

symbol	oz apoth, oz ap
system	apothecaries' unit
classification	unit of mass equal to 4.8×10^2 grams and one ounce, troy
country	United States, United Kingdom

ounce, imperial

symbol	oz
system	imperial unit
classification	unit of mass equal to 4.375×10^2 grams
country	United States, United Kingdom

ounce, troy

symbol	oz · tr
system	troy unit
classification	unit of mass equal to 4.8×10^2 grams and one ounce, apothecaries'
country	United States, United Kingdom

ounce inch squared

symbol	$oz \cdot in^2$
classification	unit of moment of inertia

ounce per cubic inch

symbol	oz/in^3
classification	unit of density (mass)

ounce per foot

symbol	oz/ft
classification	unit of density (linear)

ounce per gallon

symbol	oz/UKgal
classification	unit of density and concentration (mass) equal to 6.23602 kg/m^3
country	United Kingdom

ounce per gallon

symbol	oz/USgal
classification	unit of density and concentration (mass) equal to 7.48915 kg/m^3
country	United States

ounce per inch

symbol oz/in
classification unit of density (linear)

ounce per square foot

symbol oz/ft^2
classification unit of density (surface)

ounce per square yard

symbol oz/yd^2
classification unit of density (surface)

ounce per yard

symbol oz/yd
classification unit of density (linear)

ounce-force

symbol ozf
classification unit of force

ounce-force inch

symbol ozf · in
classification unit of moment of force and torque

ounce-force per square inch

symbol ozf/in^2
classification unit of pressure

parsec

symbol pc
system non-SI (approved)
classification unit of length equal to 3.0857×10^{16} m

pascal

symbol Pa
system SI (additional unit)
classification unit of pressure and stress

pascal cubic meter

symbol — $Pa \cdot m^3$
system — SI
classification — unit of quantity of gas

pascal cubic meter per second

symbol — $Pa \cdot m^3/s$
system — SI
classification — unit of flow rate of quantity of gas and unit of fluid escape rate

pascal liter

symbol — Pa · l, Pa · L
system — non-SI (approved)
classification — unit of quantity of gas

pascal per kelvin

symbol — Pa/K
system — SI
classification — unit of pressure coefficient

pascal per meter

symbol — Pa/m
system — SI
classification — unit of pressure gradient

pascal second

symbol — Pa · s
system — SI
classification — unit of dynamic viscosity

pascal second per meter

symbol — Pa · s/m
system — SI
classification — unit of specific impedance in acoustics and characteristic impedance of a medium

pascal second per meter cubed

symbol: $Pa \cdot s/m^3$
system: SI
classification: unit of impedance in acoustics

peck

classification: unit of volume equal to $9.09218 \times 10^{-3}\ m^3$
country: United Kingdom

peck

symbol: pk
classification: unit of volume equal to $8.80977 \times 10^{-3}\ m^3$
country: United States

pennyweight

symbol: dwt
system: troy unit
classification: unit of mass equal to 1.55517×10^{-3} kg

percent

symbol: %
classification: unit equal to 10^{-2}

per thousand

symbol: ‰
classification: unit equal to 10^{-3}

perch (not in use)

symbol: p
classification: unit of length, unit of area

phon

classification: unit of loudness level (dimensionless quantity)

phot

symbol: ph
classification: unit of illuminance

phot-second

symbol ph · s
classification unit of light exposure

pièze

symbol pz
system meter-ton-second
classification unit of pressure equal to 10^3 pascals

pint

symbol UKpt
system imperial unit
classification unit of volume equal to 3.46774×10 in^3 and 1.20095 liquid pints, used in United States

point

classification unit of mass equal to 2 milligrams

point (not in use)

classification unit of plane angle

poise

symbol P
system CGS
classification unit of dynamic viscosity

poiseuille (not in use)

symbol Pl
classification unit of dynamic viscosity

pole (not in use)

classification unit of length

poncelet (not in use)

classification French unit of power

pond

symbol p
classification unit of force

pound

symbol lb
system foot-pound-second (base unit), imperial unit and avoirdupois unit
classification unit of mass equal to 4.53592×10^{-1} kg
country United States, United Kingdom

$$\left(\text{Pound (mass)} = \frac{\text{Slug}}{\text{g}}\right)$$

pound-force

symbol lbf
system foot-pound force-second (base unit),
classification unit of force (lbf = lb (mass) × g)

pound, troy (not in use)

symbol lb · tr
system troy unit
classification unit of mass equal to 3.73242×10^{-1} kg

pound foot per second

symbol lb · ft/s
system foot-pound-second
classification unit of momentum

pound foot per second squared

symbol lb · ft/s^2
(*see* poundal)

pound foot squared

symbol lb · ft^2
system foot-pound-second
classification unit of moment of inertia

pound foot squared per second

symbol	$lb \cdot ft^2/s$
system	foot-pound-second
classification	unit of moment of momentum

pound inch squared

symbol	$lb \cdot in^2$
classification	unit of moment of inertia

pound per acre

symbol	lb/acre
classification	unit of density (surface)

pound per cubic foot

symbol	lb/ft^3
system	foot-pound-second
classification	unit of density (mass)

pound per cubic inch

symbol	lb/in^3
classification	unit of density (mass)

pound per foot

symbol	lb/ft
system	foot-pound-second
classification	unit of density (linear)

pound per foot second

symbol	$lb/(ft \cdot s)$

(*see* poundal second per square foot)

pound per gallon

symbol	lb/UKgal
classification	unit of density (mass) equal to 8.32674×10^{-1} pound/gallon, used in United States
country	United Kingdom

pound per gallon

symbol lb/USgal
classification unit of density (mass) equal to 1.20095 pounds/gallon, used in United Kingdom
country United States

pound per hour

symbol lb/h
classification unit of rate of mass flow

pound per inch

symbol lb/in
classification unit of density (linear)

pound per second

symbol lb/s
system foot-pound-second
classification unit of rate of mass flow

pound per square foot

symbol lb/ft^2
system foot-pound-second
classification unit of density (surface)

pound per square inch

symbol lb/in^2
classification unit of density (surface)

pound per square yard

symbol lb/yd^2
classification unit of density (surface)

pound per thousand square feet

symbol $lb/1000\ ft^2$
classification unit of density (surface)

pound per yard

symbol	lb/yd
classification	unit of density (linear)

poundal

symbol	pdl
system	foot-pound-second
classification	unit of force equal to 3.10810×10^{-2} pound force (poundal = pound (mass) $\times$ 1 ft/sec^2)

poundal foot

symbol	pdl · ft
system	foot-pound-second
classification	unit of moment of force and torque

poundal per square foot

symbol	pdl/ft^2
system	foot-pound-force
classification	unit of pressure

poundal second per square foot

symbol	pdl · s/ft^2
system	foot-pound-second
classification	unit of dynamic viscosity

pound-force foot

symbol	lbf · ft
system	foot-pound force-second
classification	unit of moment of force and torque

pound-force hour per square foot

symbol	lbf · h/ft^2
classification	unit of dynamic viscosity

pound-force inch

symbol	lbf · in
classification	unit of moment of force and torque

pound-force per foot

symbol lbf/ft
system foot-pound force-second
classification unit of surface tension

pound-force per inch

symbol lbf/in
classification unit of surface tension

pound-force per square foot

symbol lbf/ft^2
system foot-pound force-second
classification unit of pressure

pound-force per square inch or psia

symbol lbf/in^2
classification unit of absolute pressure that is measured with respect to zero

pound-force per square inch or psig

symbol lbf/in^2
classification unit of gauge pressure that is measured with respect to atmospheric pressure

pound-force second per square foot

symbol $lbf \cdot s/ft^2$
system foot-pound force-second
classification unit of dynamic viscosity

pound-weight (not in use)

symbol lb
classification unit same as pound-force

pour cent mille

symbol pcm
classification unit of reactivity (dimensionless quantity)

quad

classification	unit of heat energy of fuel reserves equal to 1.055×10^{18} joules
country	United States

quart

symbol	UKqt
system	imperial unit
classification	unit of volume equal to 1.13652×10^{-3} m^3
country	United Kingdom

quarter

symbol	qr
system	imperial unit and avoirdupois unit
classification	unit of mass equal to 2.8×10 pounds
country	United Kingdom

quintal

symbol	q
classification	unit of mass equal to 10^2 kg

Q-unit

classification	unit of heat energy of fuel reserves equal to 1.055×10^{21} joules

rad

symbol	rad, rd
classification	unit of energy per unit weight, specific energy, or kinetic energy per unit weight (kerma, gray, or absorbed dose)

rad per second

symbol	rad/s, rd/s
classification	unit of rate of absorbed dose, rate of kerma, rate of gray, or rate of energy per unit weight

radian

symbol	rad
system	SI (additional unit)
classification	unit of plane angle

radian per meter

symbol	rad/m
classification	unit of phase coefficient

radian per minute

symbol	rad/min
system	non-SI (approved)
classification	unit of velocity (angular)

radian per second

symbol	rad/sec
system	SI
classification	unit of velocity (angular), and of frequency (circular)

radian per second squared

symbol	rad/s^2
system	SI
classification	unit of acceleration (angular)

rayl (not in use)

classification	unit of specific impedance in acoustics

reciprocal angstrom

symbol	$Å^{-1}$
classification	unit of wavenumber

reciprocal centimeter

symbol	cm^{-1}
system	SI (multiple unit) and CGS
classification	unit of wavenumber, used in spectroscopy

reciprocal cubic meter

symbol	m^{-3}
system	SI
classification	unit of number density and molecular concentration

reciprocal cubic meter reciprocal second

symbol	$m^{-3} \cdot s^{-1}$
system	SI
classification	unit of collision rate of volume

reciprocal electronvolt reciprocal cubic meter

symbol	$eV^{-1} \cdot m^{-3}$
system	non-SI (approved)
classification	unit of density of states

reciprocal farad

symbol	F^{-1}
system	SI
classification	unit of reciprocal capacitance

reciprocal henry

symbol	H^{-1}
system	SI
classification	unit of reluctance

reciprocal joule reciprocal cubic meter

symbol	$J^{-1} \cdot m^{-3}$
system	SI
classification	unit of density of states

reciprocal kelvin

symbol	K^{-1}
system	SI
classification	unit of coefficient of linear expansion

reciprocal meter

symbol m^{-1}
system SI
classification unit of wavenumber and circular wavenumber

reciprocal minute

symbol min^{-1}
system non-SI (approved)
classification unit of frequency (circular)

reciprocal mole

symbol mol^{-1}
system SI
classification unit of Avogadro constant

reciprocal nanometer

symbol nm^{-1}
system SI (multiple unit)
classification unit of wavenumber

reciprocal ohm

symbol Ω^{-1}
(*see* siemens)

reciprocal ohm meter

symbol $1/(\Omega \cdot m)$
(*see* siemens per meter)

reciprocal pascal

symbol Pa^{-1}
system SI
classification unit of compressibility

reciprocal pascal reciprocal second

symbol $Pa^{-1} \cdot s^{-1}$
system SI
classification unit of dynamic fluidity

reciprocal poise

symbol P^{-1}
system CGS
classification unit of fluidity

reciprocal second

symbol s^{-1}
system SI
classification unit of frequency (circular)

reciprocal second reciprocal cubic meter

symbol $s^{-1} \cdot m^{-3}$
system SI
classification unit of neutron source density and decreasing density

reciprocal second reciprocal kilogram

symbol $s^{-1} \cdot kg^{-1}$
(*see* becquerel per kilogram)

reciprocal second reciprocal square meter

symbol $s^{-1} \cdot m^{-2}$
system SI
classification unit of density of molecule flow rate and rate of neutron fluence

reciprocal second reciprocal tesla

symbol $s^{-1} \cdot T^{-1}$
(*see* ampere square meter per joule second)

reciprocal square meter

symbol m^{-2}
system SI
classification unit of particle fluence

reciprocal square meter reciprocal second

symbol $m^{-2} \cdot s^{-1}$
system SI

classification unit of current density of particles, particle fluence rate, and impingement rate

register ton

classification unit of volume equal to 100 ft^3

rem (not in use)

classification unit of dose equivalent

rep (not in use)

classification unit of energy per unit of weight, specific energy, or kinetic energy per unit of weight (kerma, gray, or absorbed dose)

revolution

symbol r, rev
classification unit of plane angle

revolution per minute

symbol r/min
classification unit of frequency (rotational)

revolution per second

symbol r/s
classification unit of frequency (rotational)

reyn

(*see* poundal second per square foot)

right angle

symbol . . .^{L}
classification unit of plane angle

rod (not in use)

classification unit of length equal to 5.0292 m

Roentgen

(*see* röntgen)

röntgen

symbol	R
classification	unit of exposure

röntgen equivalent man

(*see* rem)

röntgen meter squared per curie hour

symbol	$R \cdot m^2/(Ci \cdot h)$
classification	unit of specific gamma ray constant

röntgen per second

symbol	R/s
classification	unit of rate of exposure

rood

system	imperial unit
classification	unit of area equal to 1.01171×10^3 m^2
country	United Kingdom

rutherford (not in use)

symbol	Rd
classification	unit of activity

sabin

system	foot-pound-second
classification	unit of equivalent absorption area equal to 1 ft^2

savart

classification	unit of frequency interval (dimensionless quantity)

scruple

system	apothecaries' unit
classification	unit of mass equal to 2.0×10 grams
country	United States, United Kingdom

secohm

classification	unit same as ohm second

second

symbol	s
system	SI (base unit)
classification	unit of time

second

symbol	. . ."
system	non-SI (approved)
classification	unit of plane angle

second per cubic meter

symbol	s/m^3
system	SI
classification	unit of resistance (fluid flow)

second per liter

symbol	s/l, s/L
system	non-SI (approved)
classification	unit of resistance (fluid flow)

second per meter squared

symbol	s/m^2
system	SI
classification	unit of kinematic fluidity

second squared per kilogram

symbol	s^2/kg

(*see* square meter per joule)

short hundredweight

symbol — sh cwt
system — avoirdupois unit
classification — unit of mass equal to 1.0×10^2 pounds and 8.92857×10^{-1} hundredweight (United Kingdom)
country — United States

short ton

symbol — sh tn
system — avoirdupois unit
classification — unit of mass equal to 2.0×10^3 pounds and 8.92857×10^{-1} ton (United Kingdom)
country — United States

Siegbahn unit

(*see* X-unit)

siemens

symbol — S
system — SI (additional unit)
classification — unit of admittance, modulus of admittance, conductance, susceptance

siemens meter per square millimeter

symbol — $S \cdot m/mm^2$
system — SI (multiple unit)
classification — unit of conductivity

siemens per meter

symbol — S/m
system — SI
classification — unit of conductivity

siemens square meter per mole

symbol — $S \cdot m^2/mol$
system — SI
classification — unit of molar conductivity

sievert

symbol	Sv
system	SI (additional unit)
classification	unit of dose equivalent

skot (not in use)

symbol	sk
classification	unit of scotopic luminance

slug

system	foot-pound force-second
classification	unit of mass equal to 3.21740 × 10 pounds

$$\left(\text{Slug} = \frac{\text{Pound (mass)}}{g}\right)$$

slug foot squared

symbol	slug ft^2
system	foot-pound force-second
classification	unit of moment of inertia

slug per cubic foot

symbol	slug/ft^3
system	unit of density (mass)

sone

classification	unit of loudness (dimensionless quantity)

spat

symbol	sp
classification	unit of solid angle

square centimeter

symbol	cm^2
system	SI (multiple unit) and CGS
classification	unit of area

square centimeter per dyne

symbol	cm^2/dyn
system	CGS
classification	unit of compressibility

square centimeter per erg

symbol	cm^2/erg
system	CGS
classification	unit of spectral cross section

square centimeter per kilogram-force

symbol	cm^2/kgf
classification	unit of compressibility

square centimeter per steradian erg

symbol	$cm^2/(sr \cdot erg)$
system	CGS
classification	unit of spectral angular cross section

square chain (not in use)

classification	unit of area equal to 4.84×10^2 square yards

square degree (not in use)

symbol	□°
classification	unit of solid angle

square foot

symbol	ft^2
system	imperial unit
classification	unit of area
country	United States, United Kingdom

square foot hour degree Fahrenheit per British thermal unit foot

symbol	$ft^2 \cdot h \cdot {}^\circ F/(Btu \cdot ft)$
classification	unit of thermal resistivity

square foot hour degree Fahrenheit per British thermal unit inch

symbol $ft^2 \cdot h \cdot {}^\circ F/(Btu \cdot in)$
classification unit of thermal resistivity

square foot per hour

symbol ft^2/h
(*see* foot squared per hour)

square foot per pound

symbol ft^2/lb
system foot-pound-second
classification unit of specific surface

square foot per poundal

symbol ft^2/pdl
system foot-pound-second
classification unit of compressibility

square foot per pound-force

symbol ft^2/lbf
system foot-pound force-second
classification unit of compressibility

square foot per second

symbol ft^2/s
system foot-pound-second
classification unit of thermal diffusivity

square foot per ton-force

symbol $ft^2/tonf$
classification unit of compressibility
country United Kingdom

square grade (not in use)

symbol $\square^g$
classification unit of solid angle

square inch

symbol	in^2
system	imperial unit
classification	unit of area
country	United States, United Kingdom

square inch per pound-force

symbol	in^2/lbf
classification	unit of compressibility

square inch per ton-force

symbol	$in^2/tonf$
classification	unit of compressibility
country	United Kingdom

square inch square foot

symbol	$in^2 \cdot ft^2$
classification	unit of second moment of area

square kilometer

symbol	km^2
system	SI (multiple unit)
classification	unit of area

square meter

symbol	m^2
system	SI
classification	unit of area

square meter kelvin per watt

symbol	$m^2 \cdot K/W$
system	SI
classification	unit of thermal resistance

square meter per joule

symbol	m^2/J
system	SI
classification	unit of spectral cross section

square meter per kilogram

symbol	m^2/kg
system	SI
classification	unit of mass absorption coefficient, mass attenuation coefficient, mass energy transfer coefficient, mass energy absorption coefficient, and specific surface

square meter per kilogram-force second

symbol	$m^2/(kgf \cdot s)$
system	meter-kilogram force-second
classification	unit of fluidity

square meter per mole

symbol	m^2/mol
system	SI
classification	unit of molar attenuation coefficient and molar absorption coefficient

square meter per newton

symbol m^2/N
(*see* reciprocal pascal)

square meter per newton second

symbol $m^2/(N \cdot s)$
(*see* reciprocal pascal reciprocal second)

square meter per second

symbol	m^2/s
system	SI
classification	unit of thermal diffusion coefficient, thermal diffusivity, and diffusion coefficient

square meter per steradian

symbol	m^2/sr
system	SI
classification	unit of angular cross section

square meter per steradian joule

symbol $m^2/(sr \cdot J)$
system SI
classification unit of spectral angular cross section

square meter per volt second

symbol $m^2/(V \cdot s)$
system SI
classification unit of mobility

square meter per weber

symbol m^2/Wb
(*see* square meter per volt second)

square micrometer

symbol μm^2
system SI (multiple unit)
classification unit of area

square micron

symbol μ^2
(*see* square micrometer)

square mile

symbol $mile^2$
system imperial unit
classification unit of area
country United States, United Kingdom

square mile per ton

symbol $mile^2/UKton$
classification unit of specific surface
country United Kingdom

square millimeter

symbol mm^2
system SI (multiple unit)
classification unit of area

square minute (not in use)

symbol □′
classification unit of solid angle

square second (not in use)

symbol □″
classification unit of solid angle

square yard

symbol yd^2
system imperial unit
classification unit of area
country United States, United Kingdom

square yard per ton

symbol yd^2/UKton
classification unit of specific surface
country United Kingdom

standard

classification unit of volume, used for measuring wood, equal to 1.65×10^2 cubic feet

statampere

symbol sA
system electrostatic CGS
classification unit of electric current

statampere centimeter squared

symbol $sA \cdot cm^2$
system electrostatic CGS
classification unit of electromagnetic moment

statampere per square centimeter

symbol sA/cm^2
system electrostatic CGS
classification unit of current density

statcoulomb

symbol	statC
system	electrostatic CGS
classification	unit of electric charge

statcoulomb centimeter

symbol	statC · cm
system	electrostatic CGS
classification	unit of electric dipole moment

statcoulomb per cubic centimeter

symbol	$statC/cm^3$
system	electrostatic CGS
classification	unit of density of volume of charge

statcoulomb per square centimeter

symbol	$statC/cm^2$
system	electrostatic CGS
classification	unit of electric flux density and electric polarization

statfarad

symbol	sF
system	electrostatic CGS
classification	unit of capacitance

stathenry

symbol	sH
system	electrostatic CGS
classification	unit of inductance

statmho

symbol	s℧

(*see* statsiemens)

statohm

symbol	sΩ
system	electrostatic CGS
classification	unit of resistance

statohm centimeter

symbol	sΩ · cm
system	electrostatic CGS
classification	unit of resistivity

statsiemens

symbol	sS
system	electrostatic CGS
classification	unit of resistivity

statsiemens per centimeter

symbol	sS/cm
system	electrostatic CGS
classification	unit of conductivity

stattesla

symbol	sT
classification	unit same as the electrostatic CGS unit of magnetic flux density

statvolt

symbol	sV
system	electrostatic CGS
classification	unit of electric potential

statvolt per centimeter

symbol	sV/cm
system	electrostatic CGS
classification	unit of strength of electric field

statweber

symbol	sWb
classification	unit same as the electrostatic CGS unit of magnetic flux

steradian

symbol	sr
system	SI (additional unit)
classification	unit of solid angle

stere

symbol	st
classification	unit of volume, used for measuring wood, equal to 1 m^3

sthene

symbol	sn
system	meter-ton-second
classification	unit of force equal to 10^3 newtons

sthene per square meter

symbol	sn/m^2
(*see* pièze)	

stilb

symbol	sb
classification	unit of luminance equal to one candela/square centimeter

stokes

symbol	St
system	centimeter-gram-second
classification	unit of kinematic viscosity equal to one centimeter squared per second

stone

system	imperial unit and avoirdupois unit
classification	unit of mass equal to 6.35029 kg
country	United Kingdom

survey foot

classification	unit of length equal to 1.000002 feet
country	United States

svedberg

symbol	S
classification	unit of sedimentation coefficient

talbot

classification unit of luminous energy equal to one lumen second

telegraph nautical mile (not in use)

classification unit of length equal to 6.087×10^3 feet

tesla

symbol T
system SI (additional unit)
classification unit of magnetic polarization and magnetic flux density

tesla meter

symbol $T \cdot m$
(*see* weber per meter)

tesla square meter

symbol $T \cdot m^2$
(*see* weber)

tex

symbol tex
classification unit of density (linear)

therm

classification unit of heat energy
country United Kingdom

therm per gallon

symbol therm/UKgal
classification unit of calorific value per unit of volume
country United Kingdom

thermie (not in use)

symbol th
classification unit of heat energy equal to 10^6 cal_{15}

ton

symbol UKton
system imperial unit
classification unit of mass equal to 1.12 short tons
country United Kingdom

ton

symbol USton
(*see* short ton)

ton, gross

classification unit same as ton (United Kingdom)

ton, long

classification unit same as ton (United Kingdom)

ton measurement

classification unit same as freight ton

ton, metric (not in use)

classification unit same as tonne

ton, net

classification unit same as short ton

ton, shipping

classification unit same as freight ton

ton mile

classification unit of mass carried × distance, used in traffic engineering
country United Kingdom

ton mile per gallon

symbol UKton · mile/UKgal
classification unit mass of carried × distance/volume, used in traffic engineering
country United Kingdom

ton of refrigeration

classification unit of rate of heat flow (cooling capacity)
country United States

ton (of TNT)

classification unit of energy, used in association with explosives

ton per cubic yard

symbol $UKton/yd^3$
classification unit of density (mass)
country United Kingdom

ton per hour

symbol UKton/h
classification unit of rate of mass flow
country United Kingdom

ton per mile

symbol UKton/mile
classification unit of density (linear)
country United Kingdom

ton per square mile

symbol $UKton/mile^2$
classification unit of density (surface)
country United Kingdom

ton per thousand yards

symbol UKton/1000 yd
classification unit of density (linear)
country United Kingdom

ton-force

symbol tonf
classification unit of force
country United Kingdom

ton-force foot

symbol tonf · ft
classification unit of moment of force and torque
country United Kingdom

ton-force per foot

symbol tonf/ft
classification unit of force per unit of length
country United Kingdom

ton-force per square foot

symbol tonf/ft^2
classification unit of pressure
country United Kingdom

ton-force per square inch

symbol tonf/in^2
classification unit of pressure
country United Kingdom

tonne
symbol t
system non-SI (approved) and meter-ton-second (base unit)
classification unit of mass equal to 10^3 kg

tonne kilometer

symbol t · km
system non-SI (approved)
classification unit of mass carried × distance, used in traffic engineering

tonne kilometer per liter

symbol t · km/l, t · km/L
system non-SI (approved)
classification unit of mass carried × distance/volume used in traffic engineering

tonne meter per second squared

symbol $t \cdot m/s^2$
(*see* sthene)

tonne per cubic meter

symbol t/m^3
system non-SI (approved) and meter-ton-second
classification unit of density (mass)

tonne per hectare

symbol t/ha
classification unit of density (surface)

torr

symbol Torr
classification unit of pressure

torr liter per second

symbol Torr · l/s
classification unit of fluid escape rate, used in association with vacuum measurements

tropical year

symbol a, a_{trop}
classification unit of time

var

symbol var
classification unit of reactive power equal to one watt

volt

symbol V
system SI (additional unit)
classification unit of Peltier coefficient, thermoelectromotive force, potential difference, and electric potential

volt ampere

symbol	V · A
classification	unit of apparent power equal to one watt

volt per ampere

symbol V/A
(*see* ohm)

volt per kelvin

symbol	V/K
system	SI
classification	unit of Seebeck coefficient and Thomson coefficient

volt per meter

symbol	V/m
system	SI
classification	unit of strength of electric field

volt per mil

symbol	V/mil
classification	unit of strength of electric field

volt second

symbol V · s
(*see* weber)

volt second meter

symbol V · s · m
(*see* weber meter)

volt second per ampere

symbol V · s/A
(*see* henry)

volt second per ampere meter

symbol V · s/(A · m)
(*see* henry per meter)

volt second per meter

symbol $V \cdot s/m$
(*see* weber per meter)

volt second per square meter

symbol $V \cdot s/m^2$
(*see* tesla)

volt squared per kelvin squared

symbol V^2/K^2
system SI
classification unit of Lorenz coefficient

watt

symbol W
system SI (additional unit)
classification unit of power equal to one joule per second

watt hour

symbol $W \cdot h$
system non-SI (approved)
classification unit of energy

watt per ampere squared

symbol W/A^2
(*see* ohm)

watt per cubic foot

symbol W/ft^3
classification unit of rate of heat release

watt per cubic meter

symbol W/m^3
system SI
classification unit of rate of heat release

watt per foot degree Celsius

symbol	W/(ft · °C)
classification	unit of thermal conductivity

watt per kelvin

symbol	W/K
system	SI
classification	unit of thermal conductance

watt per kilogram

symbol	W/kg
system	SI
classification	unit of rate of absorbed dose, rate of kerma, rate of gray, or rate of energy per unit of weight

watt per meter kelvin

symbol	W/(m · K)
system	SI
classification	unit of thermal conductivity

watt per square foot

symbol	W/ft^2
classification	unit of rate of heat flow

watt per square inch

symbol	W/in^2
classification	unit of rate of heat flow

watt per square meter

symbol	W/m^2
system	SI
classification	unit of density of rate of heat flow

watt per square meter kelvin

symbol	$W/(m^2 \cdot K)$
system	SI
classification	unit of coefficient of heat transfer

watt per square meter kelvin to the fourth power

symbol W/($m^2 \cdot K^4$)
system SI
classification unit of Stefan-Boltzmann constant

watt per steradian square meter

symbol W/($sr \cdot m^2$)
system SI
classification unit of radiance

watt second

symbol $W \cdot s$
(*see* joule)

watt square meter

symbol $W \cdot m^2$
system SI
classification unit of first radiation constant

weber

symbol Wb
system SI (additional unit)
classification unit of magnetic flux

weber meter

symbol $Wb \cdot m$
system SI
classification unit of magnetic dipole moment

weber per ampere

symbol Wb/A
(*see* henry)

weber per ampere meter

symbol Wb/($A \cdot m$)
(*see* henry per meter)

weber per meter

symbol	Wb/m
system	SI
classification	unit of magnetic vector potential

weber per square meter

symbol	W/m^2

(*see* tesla)

week

classification	unit of time

X-unit (not in use)

symbol	X.U.
classification	unit of wavelength

yard

symbol	yd
system	imperial unit
classification	unit of length equal to 0.9144 meter
country	United States, United Kingdom

yard per pound

symbol	yd/lb
classification	unit of specific length
country	United States, United Kingdom

year

symbol	a
classification	unit of time

6
Systems in Present Use

An enormous variety of unit systems have been used throughout human history in various parts of the world, but for scientific progress to continue it has become necessary to establish an international language, consisting of a few universally understood systems. The following basic systems are part of this modern terminology:

- CGS or centimeter-gram-second
- MKS or meter-kilogram-second
- MKfS or meter-kilogram force-second
- MKpS or meter-kilopond-second
- MKSA or meter-kilogram-second-ampere
- MTS or meter-tonne-second
- FPS or foot-pound-second
- FPfS or foot-pound force-second
- SI (base unit)
- SI (multiple unit)
- SI (additional unit)
- Non-SI (approved)
- Apothecaries' units (used in the United States and the United Kingdom)
- Avoirdupois units (used in the United States and the United Kingdom)

- Imperial units (used in the United Kingdom)
- Troy units (used in the United States and the United Kingdom)

A brief discussion of these systems follows.

Metric System

The metric system originated in 1791 when a committe of the Academy of France presented to the National Assembly a report proposing the adoption of the system. This system would later be accepted not only by the French government but also by the rest of the world, with some exceptions. It is interesting to note that two famous French scientists—J. L. Lagrange and P. S. Laplace— were members of that committee.

The new Republic of France adopted the recommendations of the committee soon afterward, in 1793. However, there was so much resistance to the gradual adoption of the new system in everyday French life that finally, in 1812 under Napoleon, the old local systems of units was officially reinstated. It was not until 1840 that the law was reversed again and use of the metric system became mandatory in the territory of France.

On March 1, 1875, the Diplomatic Conference of the Meter recommended international use of the metric system. On May 20 of the same year, seventeen nations signed the Treaty of the Meter. Then, several years later, the first General Conference on Weight and Measures was held in France and approved the new international metric prototype reference standards to redefine the units of length and mass. The Conference Generale des Poids et Mesures (CGPM) officially recognized the accuracy of the standards and distributed them among the various nations that participated in the conference.

Although the British Commonwealth steadily ignored the evolution of the metric system, the United States did legalize its use in 1866 and was among the nations that signed the Treaty of the Meter in 1875. Yet, even now use of the metric system in the United States still is not mandatory. In 1975, following the recommendations of a governmental committee, the U.S. Congress approved the Metric Conversion Act, signed by President Ford, which encourages the adoption of the metric system but does not make it mandatory.

Some confusion may arise from the similarity between kilogram (mass), kilogram (weight or force), and kilopond, included in these systems: meter-kilogram-second, meter-kilogram force-second, and meter-kilopond-second. Kilopond is, in fact, equal to kilogram force, and the term kilopond at one time was used in Central Europe as synonymous with kilogram force, but its use is now considered obsolete. Thus the two systems meter-kilogram force-second and meter-kilopond-second are synonymous.

The MKSA system referred to as the MKSA Absolute Giorgi System was originally devised in 1901 by the Italian engineer Giovanni Giorgi, was officially recognized in June 1935 in Brussels by the International Electrotechnical Commission (IEC), and was internationally confirmed in 1938. Later, in 1960, the Rationalized MKS Giorgi System constituted the basis for the SI System.

International System of Units

As the various systems of units developed over time and in different geographical areas, the need for a unified system was felt only sporadically. The constitutions of political entities, for instance, required a common system of units that would allow commercial and cultural exchanges within the boundaries of the individual states. By the late nineteenth century, at the beginning of the scientific era, visionaries were promoting worldwide acceptance to the metric system. However, only recently has the international system finally received official recognition, after far too long a period of gestation. In 1960, the Conference Generale des Poids et Mesures adopted the seven base quantities and the units indicated in Table 6-1. This system, which finally unites the nations of the world through a common scientific language, officially adopted the abbreviation SI, by which it is identified in all languages.

The Si System was attained by adopting the Rationalized MKS Giorgi System and complementing it with additional units.

SI Base Units

The Si base units, adopted in the 1960s, are shown in Table 6-1, which also shows unit nonemclature and symbols.

Table 6-1. SI base units.

Base Quantity	Base SI Unit	Symbol
length	meter	m
mass	kilogram	kg
time	second	s
electric current	ampere	A
thermodynamic temperature	kelvin	K
amount of substance	mole	mol
luminous intensity	candela	cd

Additional SI Units

In October 1980 the Comite International des Poids et Mesures added some more SI units, including two (radian and steradian) that were considered supplementary and others that were assumed to be derived. The latter ones, in fact, were obtained by dividing or multiplying the base units by the supplementary ones, or were obtained from other derived units. A list of such SI units, with nomenclature and symbols, is shown in Table 6-2.

Table 6-2. Additional SI units (supplementary and derived, adopted in 1980).

Quantity	SI Unit	Symbol
activity	becquerel	Bq
electric charge	coulomb	C
Celsius temperature	degree Celsius	°C
capacitance	farad	F
absorbed dose	gray	Gy
inductance	henry	H
frequency	hertz	Hz
energy	joule	J
luminous flux	lumen	lm

Table 6-2. *(Continued)*

Quantity	SI Unit	Symbol
illuminance	lux	lx
force	newton	N
resistance	ohm	Ω
pressure	pascal	Pa
plane angle	radian	rad
conductance	siemens	S
dose equivalent	sievert	Sv
solid angle	steradian	sr
magnetic flux density	tesla	T
electric potential	volt	V
power	watt	W
magnetic flux	weber	Wb

Non-SI Units

Outside the SI system there are many other valid units, which are in commun use in various countries. Such units, although not included in the SI units, are respected by the users of the SI system and are not controversial. A list of these units, with nomenclature and symbols, is presented in Table 6-3.

Table 6-3. Non-SI units useable in addition to the SI system.

Quantity	Unit	Symbol
time	minute	min
	hour	h
	day	d
plane angle	degree	. . .°
	minute	. . .′
	second	. . .″
volume	liter	l, L
mass	tonne	t
energy	electronvolt	eV
mass of an atom	atomic mass unit	u
length	astronomic unit	(AU)
	parsec	pc

Apothecaries' Units

The apothecaries' units derive from the seventeeth-century establishment of the pharmaceutical profession in England, and consisted of units of weight and volumes that needed to be standardized for the preparation of drugs. The system includes the following units of mass (described further in Chapter 5):

- apothecaries' ounce
- drachm
- dram
- scruple
- grain

Avoirdupois Units

The avoirdupois units, used in the United Kingdom and the United States, originated in the fourteenth century. In 1303, Edward I of England designated several units of measure that in 1335 were officially grouped and recognized under the name "avoirdupois," which in French literally means goods of weight. Constituting the system are the following units of mass (described further in Chapter 5):

- ton
- hundredweight
- cental
- quarter
- stone
- pound
- ounce
- dram
- grain
- short hundredweight
- short ton

Imperial Units

Imperial units, legally adopted in 1963, are officially valid throughout the United Kingdom. Listed below are the various units (which are described further in Chapter 5):

Units of length:
- mile
- furlong
- chain
- yard
- foot
- inch

Units of area:
- square mile
- acre
- rod
- square yard
- square foot
- square inch

Units of Volume:
- cubic yard
- cubic foot
- cubic inch

Units of capacity:
- gallon
- quart
- pint
- gill
- fluid ounce

Units of mass or weight:
- ton
- hundredweight
- cental
- quarter
- stone
- pound
- ounce
- dram
- grain

Troy Units

Troy units constitute a system for measuring mass that derived historically from Troy in France, where it was first used in the Middle Ages. The units were originally used for precious metals (gold and silver). Abolished in England in 1879, these units are no longer in use except for the ounce, its decimal parts, and multiples that are still used for measuring gold, silver, platinum, and precious stones. These units of mass include:

- troy pound
- troy ounce
- pennyweight
- grain

7 Abbreviations for Units of Measure Used in the United States in Science and Engineering

absolute	abs
acre	spell out
acre-foot	acre-ft
air horsepower	air hp
alternating-current (as adjective)	a-c
ampere	amp
ampere-hour	amp-hr
amplitude, an elliptic function	am.
Angstrom unit	Å
antilogarithm	antilog
atmosphere	atm
atomic weight	at. wt
average	avg
avoirdupois	avdp
azimuth	az or α

barometer	bar.
barrel	bbl
Baumé	Bé
board feet (feet board measure)	fbm
boiler pressure	spell out
boiling point	bp
brake horsepower	bhp
brake horsepower-hour	bhp-hr
Brinell hardness number	Bhn
British thermal unit	Btu or B
bushel	bu
calorie	cal
candle	c
candle-hour	c-hr
candlepower	cp
cent	c or ¢
center to center	c to c
centigram	cg
centiliter	cl
centimeter	cm
centimeter-gram-second (system)	cgs
chemical	chem
chemically pure	cp
circular	cir
circular mils	cir mils
coefficient	coef
cologarithm	colog
concentrate	conc
conductivity	cond
constant	const
cord	cd
cosecant	csc
cosine	cos
cosine of the amplitude, an elliptic function	cn
cost, insurance, and freight	cif
cotangent	cot
coulomb	spell out
counter electromotive force	cemf
cubic	cu

cubic centimeter	cu cm, cm^3 (liquid, meaning milliliter, ml)
cubic foot	cu ft
cubic feet per minute	cfm
cubic feet per second	cfs
cubic inch	cu in
cubic meter	cu m or m^3
cubic micron	cu μ or cu mu or μ^3
cubic millimeter	cu mm or mm^3
cubic yard	cu yd
current density	spell out
cycles per second	spell out or c
cylinder	cyl
day	spell out
decibel	db
degree	deg or °
degree centigrade	°C
degree Fahrenheit	°F
degree Kelvin	K
degree Réaumur	R
delta amplitude, an elliptic function	dn
diameter	diam
direct-current (as adjective)	d-c
dollar	$
dozen	doz
dram	dr
efficiency	eff
electric	elec
electromotive force	emf
elevation	el
equation	eq
external	ext
farad	spell out or f
feet board measure (board feet)	fbm
feet per minute	fpm
feet per second	fps
fluid	fl

foot	ft
foot-candle	ft-c
foot-Lambert	ft-L
foot-pound	ft-lb
foot-pound-second (system)	fps
foot-second (see cubic feet per second)	
franc	fr
free aboard ship	spell out
free alongside ship	spell out
free on board	fob
freezing point	fp
frequency	spell out
fusion point	fnp
gallon	gal
gallons per minute	gpm
gallons per second	gps
grain	spell out
gram	g
gram-calorie	g-cal
greatest common divisor	gcd
haversine	hav
hectare	ha
henry	h
high-pressure (adjective)	h-p
hogshead	hhd
horsepower	hp
horsepower-hour	hp-hr
hour	hr
hour (in astronomical tables)	h
hundred	C
hundredweight (112 lb)	cwt
hyperbolic cosine	cosh
hyperbolic sine	sinh
hyperbolic tangent	tanh
inch	in.
inch-pound	in-lb
inches per second	ips
indicated horsepower	ihp

indicated horsepower-hour	ihp-hr
inside diameter	ID
intermediate-pressure (adjective)	i-p
internal	int
joule	j
kilocalorie	kcal
kilocycles per second	kc
kilogram	kg
kilogram-calorie	kg-cal
kilogram-meter	kg-m
kilograms per cubic meter	kg per cu m or kg/m^3
kilograms per second	kgps
kiloliter	kl
kilometer	km
kilometers per second	kmps
kilovolt	kv
kilovolt-ampere	kva
kilowatt	kw
kilowatthour	kwhr
lambert	L
latitude	lat or ϕ
least common multiple	lcm
linear foot	lin ft
liquid	liq
lira	spell out
liter	l
logarithm (common)	log
logarithm (natural)	$\log_e$ or ln
longitude	long or λ
low-pressure (as adjective)	l-p
lumen	l
lumen-hour	l-hr
lumens per watt	lpw
mass	spell out
mathematics (ical)	math
maximum	max
mean effective pressure	mep

mean horizontal candlepower	mhcp
megacycle	spell out
megohm	spell out
melting point	mp
meter	m
meter-kilogram	m-kg
mho	spell out
microampere	μa or mu a
microfarad	μf
microinch	μin
micromicrofarad	$\mu\mu$f
micromicron	$\mu\mu$ or mu mu
micron	μ or mu
microvolt	μV
microwatt	μw or mu w
mile	spell out
miles per hour	mph
miles per hour per second	mphps
milliampere	ma
milligram	mg
millihenry	mh
millilambert	mL
milliliter	ml
millimeter	mm
millimicron	m μ or m mu
million	spell out
million gallons per day	mgd
millivolt	mV
minimum	min
minute	min
minute (angular measure)	′
minute (time) (in astronomical tables)	m
mole	spell out
molecular weight	mol.wt
month	spell out
National Electrical Code	NEC
ohm	spell out or Ω
ohm-centimeter	ohm-cm
ounce	oz

ounce-foot	oz-ft
ounce-inch	oz-in.
outside diameter	OD
parts per million	ppm
peck	pk
penny (pence)	d
pennyweight	dwt
per	
peso	spell out
pint	pt
potential	spell out
potential difference	spell out
pound	lb
pound-foot	lb-ft
pount-inch	lb-in.
pound sterling	£
pounds per brake horsepower-hour	lb per bhp-hr
pounds per cubic foot	lb per cu ft
pounds per square foot	psf
pounds per square inch	psi
pounds per square inch absolute	psia
power factor	spell out or pf
quart	qt
radian	spell out
reactive kilovolt-ampere	kvar
reactive volt-ampere	var
revolutions per minute	rpm
revolutions per second	rps
rod	spell out
root mean square	rms
secant	sec
second	sec
second (angular measure)	″
second-foot (see cubic feet per second)	
second (time) (in astronomical tables)	s
shaft horsepower	shp
shilling	s

sine	sin
sine of the ampltude, and elliptic function	sn
specific gravity	sp gr
specific heat	sp ht
spherical candle power	scp
square	sq
square centimeter	sq cm or cm^2
square foot	sq ft
square inch	sq in.
square kilometer	sq km or km^2
square meter	sq m or m^2
square micron	sq μ or sq mu or μ^2
square millimeter	sq mm or mm^2
square root of mean square	rms
standard	std.
stere	s
tangent	tan
temperature	temp
tensile strength	ts
thousand	M
thousand foot-pounds	kip-ft
thousand pounds	kip
ton	spell out
ton-mile	spell out
versed sine	vers
volt	v
volt-ampere	va
volt-coulomb	spell out
watt	w
watthour	whr
watts per candle	wpc
week	spell out
weight	wt
yard	yd
year	yr

8 The Conversion of Units

The conversion from one unit of measure to another is frequently necessary in scientific and technological fields of work. The customary way of doing this, establishing the correct mathematical proportion and calculating the result, is time consuming and unproductive compared to using a conversion factor. In the latter case, one simply multiplies the unit by a proper factor to convert the original unit into the desired one. The following table provides an alphabetical listing of conversion factors for the major units used in science and engineering. The chapter concludes with several temperature conversion tables.

To convert from	To	Multiply by
Abamperes	Amperes	10
	E.M. cgs. units of current	1
	E.S. cgs. units	2.997930×10^{10}
	Faradays (chem.)/sec.	1.036377×10^{-4}
	Faradays (phys.)/sec.	1.036086×10^{-4}
	Statamperes	2.997930×10^{10}

To convert from	To	Multiply by
Abamperes/cm.	E.M. cgs. units of surface charge density	1
	E.S. cgs. units	2.997930×10^{10}
Abamperes/sq. cm.	Amperes/circ. mil	5.0670748×10^{-5}
	Amperes/sq. cm.	10
	Amperes/sq. inch	64.516
Abampere-turns	Ampere-turns	10
Abampere-turns/cm.	Ampere-turns/cm.	10
Abcoulombs	Ampere-hours	0.0027777
	Coulombs	10
	Electronic charges	6.24196×10^{19}
	E.M. cgs. units of charge	1
	E.S. cgs. units	2.997930×10^{10}
	Faradays (chem.)	1.036377×10^{-4}
	Faradays (phys.)	1.036086×10^{-4}
	Statcoulombs	2.997930×10^{10}
Abfarads	E.M. cgs. units of capacitance	1
	E.S. cgs. units	8.987584×10^{20}
	Farads	1×10^{9}
	Microfarads	1×10^{15}
	Statfarads	8.987584×10^{20}
Abhenries	E.M. cgs. units of induction	1
	E.S. cgs. units	1.112646×10^{-21}
	Henries	1×10^{-9}
Abmhos	E.M. cgs. units of conductance	1
	E.S. cgs. units	8.987584×10^{20}
	Megamhos	1000
	Mhos	1×10^{9}
	Statmhos	8.987584×10^{20}
Abohms	E.M. cgs. units of resistance	1
	Megohms	1×10^{-15}
	Microhms	0.001
	Ohms	1×10^{-9}
	Statohms	1.112646×10^{-21}
Abohm-cm.	Circ. mil-ohms/ft.	0.0060153049
	E.M. cgs. units of resistivity	1
	Microhm-inches	0.00039370079
	Ohm-cm.	1×10^{-9}

To convert from	To	Multiply by
Abvolts	Microvolts	0.01
	Millivolts	1×10^{-5}
	Volts	1×10^{-8}
	Volts (Int.)	9.99670×10^{-9}
Abvolts/cm.	E.M. cgs. units of electric field intensity	1
	E.S. cgs. units	3.335635×10^{-11}
	Volts/cm.	1×10^{-8}
	Volts/inch	2.54×10^{-8}
	Volts/meter	1×10^{-6}
Acres	Sq. cm.	40468564
	Sq. ft.	43560
	Sq. ft. (U.S. Survey)	43559.826
	Sq. inches	6272640
	Sq. kilometers	0.0040468564
	Sq. links (Gunter's)	1×10^{5}
	Sq. meters	4046.8564
	Sq. miles (statute)	0.0015625
	Sq. perches	160
	Sq. rods	160
	Sq. yards	4840
Acre-feet	Cu. feet	43,560
	Cu. meters	1233.4818
	Cu. yards	1613.333
Acre-inches	Cu. feet	3630
	Cu. meters	102.79033
	Gallons (U.S.)	27,154.286
Amperes	Abamperes	0.1
	Amperes (Int.)	1.000165
	Cgs. units of current	1
	Mks. units of current	1
	Coulombs/sec.	1
	Coulombs (Int.)/sec.	1.000165
	Faradays (chem.)/sec.	1.036377×10^{-5}
	Faradays (phys.)/sec.	1.036086×10^{-5}
	Statamperes	2.997930×10^{9}
Amperes (Int.)	Amperes	0.999835
	Coulombs/sec.	0.999835
	Coulombs (Int.)/sec.	1

To convert from	To	Multiply by
	Faradays (chem.)/sec.	1.03623×10^{-5}
	Faradays (phys.)/sec.	1.03592×10^{-5}
Amperes/meter	Cgs. units of surface current density	0.01
	E.M. cgs. units	0.001
	E.S. cgs. units	2.997930×10^{7}
	Mks. units	1
Amperes/sq. meter	Cgs. units of volume current density	0.0001
	E.M. cgs. units	1×10^{-5}
	E.S. cgs. units	299,793.0
	Mks. units	1
Amperes/sq. mil	Abamperes/sq. cm.	15,500.031
	Amperes/sq. cm.	1.5500031×10^{5}
Ampere-hours	Abcoulombs	360
	Coulombs	3600
	Faradays (chem.)	0.373096
	Faradays (phys.)	0.372991
Ampere-turns	Cgs. units of magnetomotive force	1.2566371
	E.M. cgs. units	1.2566371
	E.S. cgs. units	3.767310×10^{10}
	Gilberts	1.2566371
Ampere-turns/weber	Cgs. units of reluctance	1.256637×10^{-8}
	E.M. cgs. units	1.256637×10^{-8}
	E.S. cgs. units	1.129413×10^{13}
	Gilberts/maxwell	1.256637×10^{-8}
Ångström units	Centimeters	1×10^{-8}
	Inches	3.9370079×10^{-9}
	Microns	0.0001
	Millimicrons	0.1
	Wavelength of orange-red line of krypton 86	0.000165076373
	Wavelength of red line of cadmium	0.000155316413
Ares	Acres	0.024710538
	Sq. dekameters	1
	Sq. feet	1076.3910
	Sq. ft. (U.S. Survey)	1076.3867

To convert from	To	Multiply by
	Sq. meters	100
	Sq. miles	3.8610216×10^{-5}
Atmospheres	Bars	1.01325
	Cm. of Hg (0°C.)	76
	Cm. of H_2O (4°C.)	1033.26
	Dynes/sq. cm.	1.01325×10^{6}
	Ft. of H_2O (39.2°F.)	33.8995
	Grams/sq. cm.	1033.23
	In. of Hg (32°F.)	29.9213
	Kg./sq. cm.	1.00323
	Mm. of Hg (0°C.)	760
	Pascals (N/sq. meter)	1.01325×10^{5}
	Pounds/sq. inch	14.6960
	Tons (short)/sq. ft.	1.05811
	Torrs	760
Atomic mass units (chem.)	Electron volts	9.31395×10^{8}
	Grams	1.66024×10^{-24}
Atomic mass units (phys.)	Electron volts	9.31141×10^{8}
	Grams	1.65979×10^{-24}
Bags (Brit.)	Bushels (Brit.)	3
Barns	Sq. cm.	1×10^{-24}
Barrels (Brit.)	Bags (Brit.)	1.5
	Barrels (U.S., dry)	1.415404
	Barrels (U.S., liq.)	1.372513
	Bushels (Brit.)	4.5
	Bushels (U.S.)	4.644253
	Cu. feet	5.779568
	Cu. meters	0.1636591
	Gallons (Brit.)	36
	Liters	163.6546
Barrels (petroleum, U.S.)	Cu. feet	5.614583
	Gallons (U.S.)	42
	Liters	158.98284
Barrels (U.S., dry)	Barrels (U.S. liq.)	0.969696
	Bushels (U.S.)	3.2812195
	Cu. feet	4.083333

To convert from	To	Multiply by
	Cu. inches	7056
	Cu. meters	0.11562712
	Quarts (U.S., dry)	105
Barrels (U.S., liq.)	Barrels (U.S., dry)	1.03125
	Barrels (wine)	1
	Cu. feet	4.2109375
	Cu. inches	7276.5
	Cu. meters	0.11924047
	Gallons (Brit.)	26.22925
	Gallons (U.S., liq.)	31.5
	Liters	119.23713
Bars	Atmospheres	0.986923
	Baryes	1×10^6
	Cm. of Hg (0°C.)	75.0062
	Dynes/sq. cm.	1×10^6
	Ft. of H_2O (60°F.)	33.4883
	Grams/sq. cm.	1019.716
	In. of Hg (32°F.)	29.5300
	Kg./sq. cm.	1.019716
	Millibars	1000
	Pounds/sq. inch	14.5038
Baryes	Atmospheres	9.86923×10^{-7}
	Bars	1×10^{-6}
	Dynes/sq. cm.	1
	Grams/sq. cm.	0.001019716
	Millibars	0.001
Bels	Decibels	10
Board feet	Cu. cm.	2359.7372
	Cu. feet	0.833333
	Cu. inches	144
Bolts of cloth	Linear feet	120
	Meters	36.576
Bougie decimales	Candies (Int.)	1.00
B.t.u.	B.t.u. (IST.)	0.999346
	B.t.u. (mean)	0.998563
	B.t.u. (39°F.)	0.994982
	B.t.u. (60°F.)	0.999689
	Calorie	251.99576
	Calorie (IST.)	251.831
	Calorie (mean)	251.634

To convert from	To	Multiply by
	Calorie (20°C.)	252.122
	Cu. cm.-atm.	10,405.6
	Ergs	1.05435×10^{10}
	Foot-poundals	25020.1
	Foot-pounds	777.649
	Gram-cm.	1.07514×10^{7}
	Hp.-hours	0.000392752
	Hp.-years	4.48347×10^{-8}
	Joules	1054.35
	Joules (Int.)	1054.18
	Kg.-meters	107.514
	Kw.-hours	0.000292875
	Kw.-hours (Int.)	0.000292827
	Liter-atm.	10.4053
	Tons of refrig. (U.S. std.)	3.46995×10^{-6}
	Watt-seconds	1054.35
	Watt-seconds (Int.)	1054.18
B.t.u. (IST.)	B.t.u.	1.00065
B.t.u. (mean)	B.t.u.	1.00144
	B.t.u. (IST.)	1.00078
	B.t.u. (39°F.)	0.996415
	B.t.u. (60°F.)	1.00113
	Hp.-hours	0.000393317
	Joules	1055.87
	Kg.-meters	107.669
	Kw.-hours	0.000293297
	Kw.-hours (Int.)	0000293248
	Liter-atm.	10.4203
	Watt-hours	0.293297
	Watt-hours (Int.)	0.293248
B.t.u. (39°F.)	B.t.u.	1.00504
	B.t.u. (IST.)	1.00439
	B.t.u. (mean)	1.00360
	B.t.u. (60°F.)	1.00473
	Joules	1059.67
B.t.u. (60°F.)	B.t.u.	1.00031
	B.t.u. (IST.)	0.999657
	B.t.u. (mean)	0.998873
	B.t.u. (39°F.)	0.995291

To convert from	To	Multiply by
B.t.u./hr.	Kilocalorie/hr.	0.251996
	Ergs/sec.	2.928751×10^6
	Foot-pounds/hr.	777.649
	Horsepower	0.000392752
	Horsepower (boiler)	2.98563×10^{-5}
	Horsepower (electric)	0.000392594
	Horsepower (metric)	0.000398199
	Kilowatts	0.000292875
	Lb. ice melted/hr.	0.0069714
	Tons of refrig. (U.S. comm.)	8.32789×10^{-5}
	Watts	0.292875
B.t.u./min.	Kilocalorie/min.	0.251996
	Ergs/sec.	1.75725×10^8
	Foot-pounds/min.	777.649
	Horsepower	0.0235651
	Horsepower (boiler)	0.00179138
	Horsepower (electric)	0.0235556
	Horsepower (metric)	0.0238920
	Joules/sec.	17.5725
	Kg.-meters/min.	107.514
	Kilowatts	0.0175725
	Lb. ice melted/hr.	0.41828
	Tons of refrig. (U.S. comm.)	0.00499673
	Watts	17.5725
B.t.u. (mean)/min.	B.t.u. (mean)/hr.	60
	Kilocalorie (mean)/hr.	15.1197
	Kilocalorie (mean)/min.	0.251996
	Ergs/sec.	1.75978×10^8
	Foot-pounds/min.	778.768
	Horsepower	0.0235990
	Horsepower (boiler)	0.00179396
	Horsepower (electric)	0.0235895
	Horsepower (metric)	0.0239264
	Joules/sec.	17.5978
	Kg.-meters/min.	107.669
	Kilowatts	0.0175978
	Lb. ice-melted/hr.	0.41888

To convert from	To	Multiply by
B.t.u./lb.	Calorie/gram	0.555555
	Cu. cm.-atm./gram	22.9405
	Cu. ft.-atm./lb.	0.367471
	Cu. ft.-(lb./sq. in.)/lb.	5.40034
	Foot-pounds/lb.	777.649
	Hp.-hr./lb.	0.000392752
	Joules/gram	2.32444
B.t.u. (mean)/lb.	Calorie (mean)/gram	0.555555
	Cu. cm.-atm./gram	22.9735
	Foot-pounds/lb.	778.768
	Hp.-hr./lb.	0.000393317
	Joules/gram	2.32779
B.t.u./sec.	B.t.u./hr.	3600
	B.t.u./min.	60
	Kilocalorie/hr.	907.185
	Kilocalorie/min.	15.1197
	Cheval-vapeur	1.43352
	Ergs/sec.	1.05435×10^{10}
	Foot-pounds/sec.	777.649
	Horsepower	1.41391
	Horsepower (boiler)	0.107483
	Horsepower (electric)	1.41334
	Horsepower (metric)	1.43352
	Kg.-meters/sec.	107.514
	Kilowatts	1.05435
	Kilowatts (Int.)	1.05418
	Watts	1054.35
	Watts (Int.)	1054.18
B.t.u. (mean)/sec.	Ergs/sec.	1.05587×10^{10}
	Foot-pounds/sec.	778.768
	Horsepower	1.41594
	Horsepower (boiler)	0.107637
	Horsepower (electric)	1.41537
	Horsepower (metric)	1.43558
	Watts	1055.87
B.t.u./sq. ft.	Calorie/sq. cm.	0.271246
B.t.u./sq. ft. × min.)	Hp./sq. ft.	0.0235651
	Kw./sq. ft.	0.0175725
	Watts/sq. in.	0.122031

To convert from	To	Multiply by
Buckets (Brit.)	Cu. cm.	18,184.35
	Gallons (Brit.)	4
Bushels (Brit.)	Bags (Brit.)	0.333333
	Bushels (U.S.)	1.032056
	Cu. cm.	36368.70
	Cu. feet	1.284348
	Cu. inches	2219.354
	Dekaliters	3.636768
	Gallons (Brit.)	8
	Hectoliters	0.3636768
	Liters	36.36768
Bushels (U.S.)	Barrels (U.S.), dry	0.3047647
	Bushels (Brit.)	0.9689395
	Cu. cm.	35,239.07
	Cu. feet	1.244456
	Cu. inches	2150.42
	Cu. meters	0.03523907
	Cu. yards	0.04609096
	Gallons (U.S., dry)	8
	Gallons (U.S., liq.)	9.309177
	Liters	35.23808
	Ounces (U.S., fluid)	1191.575
	Pecks (U.S.)	4
	Pints (U.S., dry)	64
	Quarts (U.S., dry)	32
	Quarts (U.S., liq.)	37.23671
Butts (Brit.)	Bushels (U.S.)	13.53503
	Cu. feet	16.84375
	Cu. meters	0.4769619
	Gallons (U.S.)	126
Cable lengths	Fathoms	120
	Feet	720
	Meters	219.456
Caliber	Inch	0.01
	Millimeter	0.254
Calories	B.t.u.	0.0039683207
	B.t.u. (IST.)	0.00396573
	B.t.u. (mean)	0.00396262
	B.t.u. (39°F.)	0.00394841

To convert from	To	Multiply by
	B.t.u. (60°F.)	0.00396709
	Cal. (IST.)	0.999346
	Cal. (mean)	0.998563
	Cal. (15°C.)	0.999570
	Cal. (20°C.)	1.00050
	Kilocal.	0.001
	Kilocal. (IST.)	0.000999346
	Kilocal. (mean)	0.000998563
	Kilocal. (15°C.)	0.000999570
	Kilocal. (20°C.)	0.00100050
	Cu. cm.-atm.	41.2929
	Cu. ft.-atm.	0.00145824
	Ergs	4.184×10^{7}
	Foot-poundals	99.2878
	Foot-pounds	3.08596
	Gram-cm.	42,664.9
	Hp.-hours	1.55857×10^{-6}
	Joules	4.184
	Joules (Int.)	4.18331
	Kg.-meters	0.426649
	Kw.-hours	1.162222×10^{-6}
	Liter-atm.	0.0412917
	Watt-hours	0.001162222
	Watt-hours (Int.)	0.00116203
	Watt-seconds	4.184
Calories (mean)	B.t.u.	0.00397403
	Cal.	1.00144
	Cal. (IST.)	1.00078
	Cal. (20°C.)	1.00194
	Kilocal. (mean)	0.001
	Cu. cm.-atm.	41.3523
	Cu. ft.-atm.	0.00146034
	Ergs	4.19002×10^{7}
	Foot-poundals	99.4308
	Foot-pounds	3.09040
	Hp.-hours	1.56081×10^{-6}
	Joules	4.19002
	Joules (Int.)	4.18933
	Kg.-meters	0.427263
	Kw.-hours	1.16390×10^{-6}

To convert from	To	Multiply by
	Liter-atm.	0.0413511
	Watt-seconds	4.19002
Calories (15°C.)	B.t.u.	0.00397003
	Cal.	1.00043
	Cal. (IST.)	0.999776
	Cal. (mean)	0.998992
	Cal. (20°C.)	1.00093
	Joules	4.18580
	Joules (Int.)	4.18511
Calories (20°C.)	B.t.u.	0.00396633
	Cal.	0.999498
	Cal. (IST.)	0.998845
	Cal. (mean)	0.998061
	Cal. (15°C.)	0.999068
	Joules	4.18190
	Joules (Int.)	4.18121
Cal./°C	B.t.u./°F.	0.00220462
	Joules/°F.	2.324444
	Joules (Int.)/°F.	2.32406
Cal./gram	B.t.u./lb.	1.8
	Foot-pounds/lb.	1399.77
	Joules/gram	4.184
	Watt-hours/gram	0.001162222
Cal./(gram × °C.)	B.t.u./(lb. × °C.)	1.8
	B.t.u./(lb. × °F.)	1
	Kilocal./(kg. × °C.)	1
	Joules/(gram × °C.)	4.184
	Joules/(lb. × °F.)	1054.35
Cal./hr.	B.t.u./hr.	0.0039683207
	Ergs/sec.	11,622.222
	Watts	0.001162222
Cal. (mean)/hr.	B.t.u. (mean)/hr.	0.0039683207
	Ergs/sec.	11,639.0
	Watts	0.00116390
Cal./min.	B.t.u./min.	0.0039683207
	Ergs/sec.	697,333.3
	Watts	0.069733
Cal. (mean)/min.	B.t.u. (mean)/min.	0.0039683207
	Ergs/sec.	698,337

To convert from	To	Multiply by
	Joules/sec.	0.0698337
	Watts	0.0698337
Cal./sec.	B.t.u./sec.	0.0039683207
	Ergs/sec.	4.184×10^7
	Foot-pounds/sec.	3.08596
	Horsepower	0.00561084
	Watts	4.184
Cal. (mean)/sec.	Ergs/sec.	4.19002×10^7
	Watts	4.19002
Cal./(sec. × sq. cm.)	B.t.u./(hr. × sq. ft.)	13,272.1
	Cal./(hr. × sq. cm.)	3600
	Watts/sq. cm.	4.184
Cal./(sec. × sq. cm. × °C.)	B.t.u./(hr. × sq. ft. × °F.)	7373.38
Cal./sq. cm.	B.t.u./sq. ft.	3.68669
$\frac{\text{Cal.-cm.}}{(\text{hr.} \times \text{sq. cm.} \times {}^\circ\text{C.})}$	$\frac{\text{B.t.u.-ft.}}{(\text{hr.} \times \text{sq. ft.} \times {}^\circ\text{F.})}$	0.0671969
	$\frac{\text{B.t.u.-inch}}{(\text{hr.} \times \text{sq. ft.} \times {}^\circ\text{F.})}$	0.806363
Cal.-cm./sq. cm.	B.t.u.-inch/sq. ft.	1.4514530
Cal.-sec.	Planck's constant	6.31531×10^{22}
Cal.-sec./Avog. No. (chem.)	Planck's constant	1.04849×10^{10}
Cal.-sec./Avog. No. (phys.)	Planck's constant	1.04821×10^{10}
*Candles (English)	Candles (Int.)	1.04
	Hefner units	1.16
Candles (German)	Candles (English)	1.01
	Candles (Int.)	1.05
	Hefner units	1.17
Candles (Int.)	Candles (English)	0.96
	Candles (German)	0.95
	Candles (pentane)	1.00
	Hefner units	1.11
	Lumens (Int.)/steradian	1

*Candle is equivalent to candela (SI unit of luminous intensity)

To convert from	To	Multiply by
Candles (pentane)	Candles (Int.)	1.00
Candles/sq. cm.	Candles/sq. inch	6.4516
	Candles/sq. meter	10000
	Foot-lamberts	2918.6351
	Lamberts	3.1415927
Candles/sq. ft.	Candles/sq. inch	0.0069444
	Candles/sq. meter	10.763910
	Foot-lamberts	3.1415927
	Lamberts	0.0033815822
Candles/sq. inch	Candles/sq. cm.	0.15500031
	Candles/sq. foot	144
	Foot-lamberts	452.38934
	Lamberts	0.48694784
Candle power (spher.)	Lumens	12.566370
Carats (parts of gold per 24 of mixture)	Milligrams/gram	41.6666
Carats (1877)	Grains	3.168
	Milligrams	205.3
Carats (metric)	Grains	3.08647
	Grams	0.2
	Milligrams	200
Carcel units	Candles (Int.)	9.61
Centals	Kilograms	45.359237
	Pounds	100
Centares	Ares	0.01
	Sq. feet	10.763910
	Sq. inches	1550.0031
	Sq. meters	1
	Sq. yards	1.1959900
Centigrams	Grains	0.15432358
	Grams	0.01
Centiliters	Cu. cm.	10.00028
	Cu. inches	0.6102545
	Liters	0.01
	Ounces (U.S., fluid)	0.3381497
Centimeters	Ångström units	1×10^8
	Feet	0.032808399
	Feet (U.S. Survey)	0.032808333
	Hands	0.098425197

To convert from	To	Multiply by
	Inches	0.39370079
	Links (Gunter's)	0.049709695
	Links (Ramden's)	0.032808399
	Meters	0.01
	Microns	10,000
	Miles (naut., Int.)	5.3995680×10^{-6}
	Miles (statute)	6.2137119×10^{-6}
	Millimeters	10
	Millimicrons	1×10^{7}
	Mils	393.70079
	Picas (printer's)	2.3710630
	Points (printer's)	28.452756
	Rods	0.0019883878
	Wavelength of orange-red line of krypton 86	16,507.6373
	Wavelength of red line of cadmium	15,531.6413
	Yards	0.010936133
Cm. of Hg (0°C.)	Atmospheres	0.013157895
	Bars	0.0133322
	Dynes/sq. cm.	13,332.2
	Ft. of H_2O (4°C.)	0.446050
	Ft. of H_2O (60°F.)	0.446474
	In. of Hg (0°C.)	0.39370079
	Kg./sq. meter	135.951
	Pounds/sq. ft.	27.8450
	Pounds/sq. inch	0.193368
	Torrs	10
Cm. of H_2O (4°C.)	Atmospheres	0.000967814
	Dynes/sq. cm.	980.638
	Pounds/sq. inch	0.0142229
Centimeters/sec.	Feet/min.	1.9685039
	Feet/sec.	0.032808399
	Kilometers/hr.	0.036
	Kilometers/min.	0.0006
	Knots (Int.)	0.019438445
	Meters/min.	0.6
	Miles/hr.	0.022369363
	Miles/min.	0.00037282272

To convert from	To	Multiply by
Cm./(sec. × sec.)	Kilometers/(hr. × sec.)	0.036
	Miles/(hr. × sec.)	0.022369363
Centimeters/year	Inches/year	0.39370079
Centipoises	Grams/(cm. × sec.)	0.01
	Poises	0.01
	Pound/(ft. × hr.)	2.4190883
	Pounds/(ft. × sec.)	0.00067196898
Centistokes	Stokes	0.01
Chains (Gunter's)	Centimeters	2011.68
	Chains (Ramden's)	0.66
	Feet	66
	Feet (U.S. Survey)	65.999868
	Furlongs	0.1
	Inches	792
	Links (Gunter's)	100
	Links (Ramden's)	66
	Meters	20.1168
	Miles (statute)	0.0125
	Rods	4
	Yards	22
Chains (Ramden's)	Centimeters	3048
	Chains (Gunter's)	1.515151
	Feet	100
	Feet (U.S. Survey)	99.999800
Cheval-vapeur	Horsepower (metric)	1
Cheval-vapeur-heures	Joules	2,647,795
Circles	Degrees	360
	Grades	400
	Minutes	21,600
	Radians	6.2831853
	Signs	12
Circular inches	Circular mm.	645.16
	Sq. cm.	5.0670748
	Sq. inches	0.78539816
Circular mm.	Sq. cm.	0.0078539816
	Sq. inches	0.0012173696
	Sq. mm.	0.78539816

To convert from	To	Multiply by
Circular mils	Circular inches	1×10^{-6}
	Sq. cm.	5.0670748×10^{-6}
	Sq. inches	$7.8539816 + 10^{-7}$
	Sq. mm.	0.00050670748
	Sq. mils	0.78539816
Circumferences	Degrees	360
	Grades	400
	Minutes	21,600
	Radians	6.2831853
	Seconds	1,296,000
Cords	Cord-feet	8
	Cu. feet	128
	Cu. meters	3.6245734
Cord-feet	Cords	0.125
	Cu. feet	16
Coulombs	Abcoulombs	1
	Ampere-hours	0.0002777
	Ampere-seconds	1
	Coulombs (Int.)	1.000165
	Electronic charge	6.24196×10^{18}
	E.M. cgs. units of electric charge	0.1
	E.S. cgs. units of electric charge	2.997930×10^{9}
	Faradays (chem.)	1.036377×10^{-5}
	Faradays (phys.)	1.036086×10^{-5}
	Mks. units of electric charge	1
	Statcoulombs	2.997930×10^{9}
Coulombs/cu. meter	E.M. cgs. units of volume charge density	1×10^{-7}
	E.S. cgs. units	2997.930
Coulombs/sq. cm.	Abcoulombs/sq. cm.	0.1
	Cgs. units of polarization, and surface charge density	1
Cubic centimeters	Board feet	0.00042377600
	Bushels (Brit.)	2.749617×10^{-5}
	Bushels (U.S.)	2.837759×10^{-5}
	Cu. feet	3.5314667×10^{-5}

To convert from	To	Multiply by
	Cu. inches	0.061023744
	Cu. meters	1×10^{-6}
	Cu. yards	1.3079506×10^{-6}
	Drachms (Brit., fluid)	0.28156080
	Drams (U.S., fluid)	0.27051218
	Gallons (Brit.)	0.0002199694
	Gallons (U.S., dry)	0.00022702075
	Gallons (U.S., liq.)	0.00026417205
	Gills (Brit.)	0.007039020
	Gills (U.S.)	0.0084535058
	Liters	0.000999972
	Ounces (Brit., fluid)	0.03519510
	Ounces (U.S., fluid)	0.033814023
	Pints (U.S., dry)	0.0018161660
	Pints (U.S., liq.)	0.0021133764
	Quarts (Brit.)	0.0008798775
	Quarts (U.S., dry)	0.00090808298
	Quarts (U.S., liq.)	0.0010566882
Cu. cm./gram	Cu. ft./lb.	0.016018463
Cu. cm./sec.	Cu. ft./min.	0.0021188800
	Cal. (U.S.)/min.	0.015850323
	Gal. (U.S.)/sec.	0.00026417205
Cu. cm.-atm.	B.t.u.	9.61019×10^{-5}
	B.t.u. (mean)	9.59637×10^{-5}
	Cal.	0.0242173
	Cal. (mean)	0.0241824
	Cu.-ft.-atm.	3.5314667×10^{-5}
	Joules	0.101325
	Watt-hours	2.81458×10^{-5}
Cu. cm.-atm./gram.	B.t.u./lb.	0.0435911
	Cal./gram	0.0242173
	Cu. ft.-(lb./sq. in.)/lb.	0.235406
	Ft.-lb./lb.	33.8985
	Joules/gram	0.101325
	Kg.-meters/gram	0.0103323
	Kw.-hr./gram	2.81458×10^{-8}
Cubic decimeters	Cu. cm.	1000
	Cu. feet	0.035316667
	Cu. inches	61.023744

To convert from	To	Multiply by
	Cu. meters	0.001
	Cu. yards	0.0013079506
	Liters	0.999972
Cubic dekameters	Cu. decimeters	1×10^6
	Cu. feet	35,314.667
	Cu. inches	6.1023744×10^7
	Cu. meters	1000
	Liters	999,972
Cubic feet	Acre-feet	2.2956841×10^{-5}
	Board feet	12
	Bushels (Brit.)	0.7786049
	Bushels (U.S.)	0.80356395
	Cords (wood)	0.0078125
	Cord-feet	0.0625
	Cu. centimeters	28,316.847
	Cu. meters	0.028316847
	Gallons (U.S., dry)	6.4285116
	Gallons (U.S., liq.)	7.4805195
	Liters	28.31605
	Ounces (Brit., fluid)	996.6143
	Ounces (U.S., fluid)	957.50649
	Pints (U.S., liq.)	59.844156
	Quarts (U.S., dry)	25.714047
	Quarts (U.S., liq.)	29.922078
Cu. ft. of H_2O (39.2°F.)	Pounds of H_2O	62.4262
Cu. ft. of H_2O (60°F.)	Pounds of H_2O	63.3663
Cu. ft./hr.	Acre-feet/hr.	2.2956841×10^{-5}
	Cu. cm./sec.	7.8657907
	Cu. ft./day	24
	Gal. (U.S.)/hr.	7.4805195
	Liters/hr.	28.31605
Cu. ft./min.	Acre-feet/hr.	0.0013774105
	Acre-feet/min.	2.2956841×10^{-5}
	Cu. cm./sec.	471.94744
	Cu. ft./hr.	60
	Gal. (U.S.)/min.	7.4805195
	Liters/sec.	0.4719342

To convert from	To	Multiply by
Cu. ft./lb.	Cu. cm./gram	62.427961
	Millimeters/gram	62.42621
Cu. ft./sec.	Acre-inches/hr.	0.99173553
	Cu. cm./sec.	28,316.847
	Cu. yards/min.	2.222222
	Gal. (U.S.)/min.	448.83117
	Liters/min.	1698.963
	Liters/sec.	28.31605
Cu. ft. of H_2O (60°F.)/sec.	Lb. of H_2O/min.	3741.98
Cu. ft.-atm.	B.t.u.	2.72130
	Cal.	685.756
	Cu. cm.-atm.	28,316.847
	Cu. ft.-(lb/sq. in.)	14.6960
	Foot-pounds	2116.22
	Hp.-hours	0.00106880
	Joules	2869.20
	Kg.-meters	292.577
	Kw.-hours	0.000797001
Cubic inches	Barrels (Brit.)	0.0001001292
	Barrels (U.S., dry)	0.00014172336
	Board feet	0.0069444
	Bushels (Brit.)	0.0004505815
	Bushels (U.S.)	0.00046502544
	Cu. cm.	16.387064
	Cu. feet	0.00057870370
	Cu. meters	1.6387064×10^{-5}
	Cu. yards	2.1433470×10^{-5}
	Drams (U.S., fluid)	4.4329004
	Gallons (Brit.)	0.003604652
	Gallons (U.S., dry)	0.0037202035
	Gallons (U.S., liq.)	0.0043290043
	Liters	0.01638661
	Milliliters	16.38661
	Ounces (Brit., fluid)	0.5767444
	Ounces (U.S., fluid)	0.55411255
	Pecks (U.S.)	0.0018601017
	Pints (U.S., dry)	0.029761628
	Pints (U.S., liq.)	0.034632035

To convert from	To	Multiply by
	Quarts (U.S., dry)	0.014880814
	Quarts (U.S., liq.)	0.017316017
Cu. in. of H_2O (4°C.)	Pounds of H_2O	0.0361263
Cu. in. of H_2O (60°F.)	Pounds of H_2O	0.0360916
Cubic meters	Acre-feet	0.00081071319
	Barrels (Brit.)	6.110261
	Barrels (U.S., dry)	8.648490
	Barrels (U.S., liq.)	8.3864145
	Bushels (Brit.)	27.49617
	Bushels (U.S.)	28.377593
	Cu. cm.	1×10^6
	Cu. feet	35.314667
	Cu. inches	61,023.74
	Cu. yards	1.3079506
	Gallons (Brit.)	219.9694
	Gallons (U.S., liq.)	264.17205
	Hogshead	4.1932072
	Liters	999.972
	Pints (U.S., liq.)	2113.3764
	Quarts (U.S., liq.)	1056.6882
	Steres	1
Cu. meters/min.	Gal. (Brit.)/min.	219.9694
	Gal. (U.S.)/min.	264.1721
	Liters/min.	999.972
Cu. millimeters	Cu. cm.	0.001
	Cu. inches	6.1023744×10^{-1}
	Cu. meters	1×10^{-9}
	Minims (Brit.)	0.01689365
	Minims (U.S.)	0.016230731
Cu. yards	Bushels (Brit.)	21.02233
	Bushels (U.S.)	21.696227
	Cu. cm.	764,554.86
	Cu. feet	27
	Cu. inches	46.656
	Cu. meters	0.76455486
	Gallons (Brit.)	168.1787
	Gallons (U.S., dry)	173.56981
	Gallons (U.S., liq.)	201.97403

To convert from	To	Multiply by
	Liters	764.5335
	Quarts (Brit.)	672.7146
	Quarts (U.S., dry)	694.27926
	Quarts (U.S., liq.)	807.89610
Cu. yd./min.	Cu. ft./sec.	0.45
	Gal. (U.S.)/sec.	3.3662338
	Liters/sec.	12.74222
Cubits	Centimeters	45.72
	Feet	1.5
	Inches	18
Daltons (chem.)	Grams	1.66024×10^{-24}
Daltons (phys.)	Grams	1.65979×10^{-24}
Days (mean solar)	Days (sidereal)	1.00273791
	Hours (mean solar)	24
	Hours (sidereal)	24.065710
	Years (calendar)	0.0027397260
	Years (sidereal)	0.0027378031
	Years (tropical)	0.0027379093
Days (sidereal)	Days (mean solar)	0.99726957
	Hours (mean solar)	23.934470
	Hours (sidereal)	24
	Minutes (mean solar)	1436.0682
	Minute (sidereal)	1440
	Second (sidereal)	86,400
	Years (calendar)	0.0027322454
	Years (sidereal)	0.0027303277
	Years (tropical)	0.0027304336
Decibels	Bels	0.1
Decimeters	Centimeters	0.1
	Feet	0.32808399
	Feet (U.S. Survey)	0.328083333
	Inches	3.9370079
	Meters	0.1
Decisteres	Cu. meters	0.1
Degrees	Circles	0.0027777
	Minutes	60
	Quadrants	0.0111111
	Radians	0.017453293
	Seconds	3600

To convert from	To	Multiply by
Degrees/cm.	Radians/cm.	0.017453293
Degrees/foot	Radians/cm.	0.00057261458
Degrees/inch	Radian/cm.	0.0068713750
Degrees/min.	Degrees/sec.	0.0166666
	Radians/sec.	0.00029088821
	Revolutions/sec.	4.629629×10^{-5}
Degrees/sec.	Radians/sec.	0.017453293
	Revolutions/min.	0.166666
	Revolutions/sec.	0.0027777
Dekaliters	Pecks (U.S.)	1.135136
	Pints (U.S., dry)	18.16217
Dekameters	Centimeters	1000
	Feet	32.808399
	Feet (U.S. Survey)	32.808333
	Inches	393.70079
	Kilometers	0.01
	Meters	10
	Yards	10.93613
Demals	Gram-equiv./cu. decimeter	1
Drachms (Brit. fluid)	Cu. cm.	3.551631
	Cu. inches	0.2167338
	Drams (U.S., fluid)	0.9607594
	Milliliters	3.551531
Drams (apoth. or troy)	Drams (avdp.)	2.1942857
	Grains	60
	Grams	3887.9346
	Ounces (apoth. *or* troy)	0.125
	Ounces (avdp.)	0.13714286
	Scruples (apoth.)	3
Drams (avdp.)	Drams (apoth. *or* troy)	0.455729166
	Grains	27.34375
	Grams	1.7718452
	Ounces (apoth. *or* troy)	0.056966146
	Ounces (avdp.)	0.0625
	Pennyweights	1.1393229
	Pounds (apoth. *or* troy)	0.0047471788
	Pounds (avdp.)	0.00390625
	Scruples (apoth.)	1.3671875

To convert from	To	Multiply by
Drams (U.S., fluid)	Cu. cm.	3.6967162
	Cu. inches	0.22558594
	Drachms (Brit., fluid)	1.040843
	Gills (U.S.)	0.03125
	Milliliters	3.696588
	Minims (U.S.)	60
	Ounces (U.S., fluid)	0.125
	Pints (U.S., liq.)	0.0078125
Dynes	Grains	0.01573663
	Grams	0.001019716
	Newtons	0.00001
	Poundals	7.2330138×10^{-5}
	Pounds	2.248089×10^{-6}
Dynes/cm.	Ergs/sq. cm.	1
	Ergs/sq. mm.	0.01
	Grams/cm.	0.001019716
	Poundals/inch	0.00018371855
Dynes/cu. cm.	Grams/cu. cm.	0.001019716
	Poundals/cu. inch	0.0011852786
Dynes/sq. cm.	Atmospheres	9.86923×10^{-7}
	Bars	1×10^{-6}
	Baryes	1
	Cm. of Hg (0°C.)	7.50062×10^{-5}
	Cm. of H_2O (4°C.)	0.001019745
	Grams/sq. cm.	0.001019716
	In. of Hg (32°F.)	2.95300×10^{-5}
	In. of H_2O (4°C.)	0.000401474
	Kg./sq. meter	0.01019716
	Pascals (N/sq. meter)	0.1
	Poundals/sq. in.	0.00046664510
	Pounds/sq. in.	1.450377×10^{-5}
Dyne-centimeters	Ergs	1
	Foot-poundals	2.3730360×10^{-6}
	Foot-pounds	7.37562×10^{-8}
	Gram-cm.	0.001019716
	Inch-pounds	8.85075×10^{-7}
	Kg.-meters	1.019716×10^{-8}
	Newton-meters	1×10^{-7}

To convert from	To	Multiply by
Electron volts	Ergs	1.60209×10^{-12}
	Grams	1.78253×10^{-33}
Electronic charges	Abcoulombs	1.60209×10^{-20}
	Coulombs	1.60209×10^{-19}
	Statcoulombs	4.80296×10^{-10}
Electronic charges/kg.	Statcoulombs/dyne	4.89766×10^{-16}
E.S. cgs. units of induction flux	E.M. cgs. units	2.997930×10^{10}
E.S. cgs. units of magnetic charge	E.M. cgs. units	2.997930×10^{10}
E.S. cgs. units of magnetic field intensity	E.M. cgs. units	3.335635×10^{-11}
Ells	Centimeters	114.3
	Inches	45
Ergs	B.t.u.	9.48451×10^{-11}
	Cal.	2.39006×10^{-8}
	Kilocal.	2.39006×10^{-11}
	Kilocal. (20°C.)	2.39126×10^{-11}
	Cu. cm.-atm.	9.86923×10^{-7}
	Cu. ft.-atm.	3.48529×10^{-11}
	Cu. ft.-(lb./sq. in.)	5.12196×10^{-10}
	Dyne-cm.	1
	Electron volts	6.24196×10^{11}
	Foot-poundals	2.3730360×10^{-6}
	Foot-pounds	7.37562×10^{-8}
	Gram-cm.	0.001019716
	Joules	1×10^{-7}
	Joules (Int.)	9.99835×10^{-8}
	Kw.-hours	2.777777×10^{-14}
	Kg.-meters	1.019716×10^{-8}
	Liter-atm.	9.86895×10^{-10}
	Watt-sec.	1×10^{-7}
Ergs/(gram-mol. × °C.)	Foot-pounds/(lb.-mol. × °F.)	1.85863×10^{-5}
Ergs/sec.	B.t.u./min.	5.69071×10^{-9}
	Cal./min.	1.43403×10^{-6}
	Dyne-cm./sec.	1

To convert from	To	Multiply by
	Foot-pounds/min.	4.42537×10^{-6}
	Gram-cm./sec.	0.001019716
	Horsepower	1.34102×10^{-10}
	Joules/sec.	1×10^{-7}
	Kilowatts	1×10^{-10}
	Watts	1×10^{-7}
Ergs/sq. cm.	Dynes/cm.	1
	Ergs/sq. mm.	0.01
Ergs/sq. mm.	Dynes/cm.	100
	Ergs/sq. cm.	100
Erg-sec.	Planck's constant	1.50932×10^{26}
Farads	Abfarads	1×10^{-9}
	E.M. cgs. units	1×10^{-9}
	E.S. cgs. units	8.987584×10^{11}
	Farads (Int.)	1.000495
	Microfarads	1×10^{6}
	Statfarads	8.98758×10^{11}
Farads (Int.)	Farads	0.999505
Fathoms	Centimeters	182.88
	Feet	6
	Inches	72
	Meters	1.8288
	Miles (naut., Int.)	0.00098747300
	Miles (statute)	0.001136363
	Yards	2
Feet	Centimeters	30.48
	Chains (Gunter's)	0.01515151
	Fathoms	0.166666
	Feet (U.S. Survey)	0.99999800
	Furlongs	0.00151515
	Inches	12
	Meters	0.3048
	Microns	304,800
	Miles (naut., Int.)	0.00016457883
	Miles (statute)	0.000189393
	Rods	0.060606
	Ropes (Brit.)	0.05
	Yards	0.333333

To convert from	To	Multiply by
Feet (U.S. Survey)	Centimeters	30.480061
	Chains (Gunter's)	0.015151545
	Chains (Ramden's)	0.010000020
	Feet	1.0000020
	Inches	12.000024
	Links (Gunter's)	1.5151545
	Links (Ramden's)	1.0000020
	Meters	0.30480061
	Miles (statute)	0.00018939432
	Rods	0.060606182
	Yards	0.33333400
Feet of air (1 atm., 60°F.)	Atmospheres	3.6083×10^{-5}
	Ft. of Hg (32°F.)	0.00089970
	Ft. of H_2O (60°F.)	0.0012244
	In. of Hg (32°F.)	0.0010796
	Pounds/sq. inch	0.00053027
Feet of Hg (32°F.)	Cm. of Hg (0°C.)	30.48
	Ft. of H_2O (60°F.)	13.6085
	In. of H_2O (60°F.)	163.302
	Ounces/sq. inch	94.3016
	Pounds/sq. inch	5.89385
Feet of H_2O (4°C.)	Atmospheres	0.0294990
	Cm. of Hg (0°C.)	2.24192
	Dynes/sq. cm.	29889.8
	Grams/sq. cm.	30.4791
	In. of Hg (32°F.)	0.882646
	Kg./sq. meter	304.791
	Pascals (N/sq. meter)	2989.07
	Pounds/sq. inch	0.433515
Feet/hour	Cm./hr.	30.48
	Cm./min.	0.508
	Cm./sec.	0.0084666
	Feet/min.	0.0166666
	Inches/hr.	12
	Kilometers/hr.	0.0003048
	Kilometers/min.	5.08×10^{-6}
	Knots (Int.)	0.0001645788
	Miles/hr.	0.000189393

To convert from	To	Multiply by
	Miles/min.	3.156565×10^{-6}
	Miles/sec.	5.2609428×10^{-8}
Feet/minute	Cm./sec.	0.508
	Feet/sec.	0.0166666
	Kilometers/hr.	0.018288
	Meters/min.	0.3048
	Meters/sec.	0.00508
	Miles/hr.	0.01136363
Feet/second	Cm./sec.	30.48
	Kilometers/hr.	1.09728
	Kilometers/min.	0.018288
	Meters/min.	18.288
	Miles/hr.	0.68181818
	Miles/min.	0.01136363
Feet/(sec. × sec.)	Kilometers/(hr. × sec.)	1.09728
	Meters/(sec. × sec.)	0.3048
	Miles/(hr. × sec.)	0.68181818
Firkins (Brit.)	Bushels (Brit.)	1.125
	Cu. cm.	40914.79
	Cu. feet	1.444892
	Firkins (U.S.)	1.200949
	Gallons (Brit.)	9
	Liters	40.91364
	Pints (Brit.)	72
Firkins (U.S.)	Barrels (U.S., dry)	0.29464286
	Barrels (U.S., liq.)	0.28571429
	Bushels (U.S.)	0.96678788
	Cu. feet	1.203125
	Firkins (Brit.)	0.8326747
	Liters	34.06775
	Pints (U.S., liq.)	72
Foot-candles	Lumens/sq. ft.	1
	Lumens/sq. meter	10.763910
	Lux	10.763910
	Milliphots	1.0763910
Foot-lamberts	Candles/sq. cm.	0.00034262591
	Candles/sq. ft.	0.31830989
	Millilamberts	1.0763910
	Lamberts	0.0010763910
	Lumens/sq. ft.	1

To convert from	To	Multiply by
Foot-poundals	B.t.u.	3.99678×10^{-5}
	B.t.u. (IST.)	3.99417×10^{-5}
	B.t.u. (mean)	3.99104×10^{-5}
	Cal.	0.0100717
	Cal. (IST.)	0.0100651
	Cal. (mean)	0.0100573
	Cu. cm.-atm.	0.415890
	Cu. ft.-atm.	1.46870×10^{-5}
	Dyne-cm.	4.2140110×10^{5}
	Ergs	4.2140110×10^{5}
	Foot-pounds	0.0310810
	Hp.-hours	1.56974×10^{-8}
	Joules	0.042140110
	Joules (Int.)	0.0421332
	Kg.-meters	0.00429710
	Kw.-hours	1.17056×10^{-8}
	Liter-atm.	0.000415879
	B.t.u.	0.00128593
Foot-pounds	B.t.u. (IST.)	0.00128509
	B.t.u. (mean)	0.00128408
	Cal.	0.324048
	Cal. (IST.)	0.323836
	Cal. (mean)	0.323582
	Cal. (20°C.)	0.324211
	Kilocal.	0.000324048
	Kilocal. (IST.)	0.000323836
	Kilocal. (mean)	0.000323582
	Cu. ft.-atm.	0.000472541
	Dyne-cm.	1.35582×10^{7}
	Ergs	1.35582×10^{7}
	Foot-poundals	32.1740
	Gram-cm.	13,825.5
	Hp.-hours	5.05050×10^{-7}
	Joules	1.35582
	Kg.-meters	0.138255
	Kw.-hours	3.76616×10^{-7}
	Kw.-hours (Int.)	3.76554×10^{-7}
	Liter-atm.	0.0133805
	Newton-meters	1.3558180

To convert from	To	Multiply by
	Lb. H_2O evap. from and at 212°F.	1.3245×10^{-6}
	Watt-hours	0.000376616
Foot-pounds/hr.	B.t.u./min.	2.14321×10^{-5}
	B.t.u. (mean)/min.	2.14013×10^{-5}
	Cal./min.	0.00540080
	Cal. (mean)/min.	0.00539304
	Ergs/min.	2.25970×10^{5}
	Foot-pounds/min.	0.0166666
	Horsepower	5.050505×10^{-7}
	Horsepower (metric)	5.12055×10^{-7}
	Kilowatts	3.76616×10^{-7}
	Watts	0.000376616
	Watts (Int.)	0.000376554
Foot-pounds/min.	B.t.u./sec.	2.14321×10^{-5}
	B.t.u. (mean)/sec.	2.14013×10^{-5}
	Cal./sec.	0.00540080
	Cal. (mean)/sec.	0.00539304
	Ergs/sec.	2.25970×10^{5}
	Foot-pounds/sec.	0.0166666
	Horsepower	3.030303×10^{-5}
	Horsepower (metric)	3.07233×10^{-5}
	Joules/sec.	0.0225970
	Joules (Int.)/sec.	0.0225932
	Kilowatts	2.25970×10^{-5}
	Watts	0.0225970
Foot-pounds/lb.	B.t.u./lb.	0.00128593
	B.t.u. (IST.)/lb.	0.00128509
	B.t.u. (mean)/lb.	0.00128408
	Cal/gm.	0.000714404
	Cal. (IST.)/gram	0.000713937
	Cal. (mean)/gram	0.000713377
	Hp.-hr./lb.	5.05050×10^{-7}
	Joules/gram	0.00298907
	Kg.-meters/gram	0.000304800
	Kw.-hr./gram	8.30296×10^{-10}
Foot-pounds/sec.	B.t.u./min.	0.0771556
	B.t.u. (mean)/min.	0.0770447
	B.t.u./sec.	0.00128593

To convert from	To	Multiply by
	B.t.u. (mean)/sec.	0.00128408
	Cal./sec.	0.324048
	Cal. (mean)/sec.	0.323582
	Ergs/sec.	1.35582×10^7
	Gram-cm./sec.	13,825.5
	Horsepower	0.00181818
	Joules/sec.	1.35582
	Kilowatts	0.00135582
	Watts	1.35582
	Watts (Int.)	1.35559
Furlongs	Centimeters	20,116.8
	Chains (Gunter's)	10
	Chains (Ramden's)	6.6
	Feet	660
	Inches	7920
	Meters	201.168
	Miles (naut., Int.)	0.10862203
	Miles (statute)	0.125
	Rods	40
	Yards	220
Gallons (Brit.)	Barrels (Brit.)	0.027777
	Bushels (Brit.)	0.125
	Cu. centimeters	4546.087
	Cu. feet	0.1605436
	Cu. inches	277.4193
	Drachms (Brit. fluid)	1280
	Firkins (Brit.)	0.111111
	Gallons (U.S., liq.)	1.200949
	Gills (Brit.)	32
	Liters	4.545960
	Minims (Brit.)	76,800
	Ounces (Brit., fluid)	160
	Ounces (U.S., fluid)	153.7215
	Pecks (Brit.)	0.5
	Lb. of H_2O (62°F.)	10
Gallons (U.S., dry)	Barrels (U.S., dry)	0.038095592
	Barrels (U.S., liq.)	0.036941181
	Bushels (U.S.)	0.125

To convert from	To	Multiply by
	Cu. centimeters	4404.8828
	Cu. feet	0.15555700
	Cu. inches	268.8025
	Gallons (U.S., liq.)	1.16364719
	Liters	4.404760
Gallons (U.S., liq.)	Acre-feet	3.0688833×10^{-6}
	Barrels (U.S., liq.)	0.031746032
	Barrels (petroleum, U.S.)	0.023809524
	Bushels (U.S.)	0.10742088
	Cu. centimeters	3785.4118
	Cu. feet	0.133680555
	Cu. inches	231
	Cu. meters	0.0037854118
	Cu. yards	0.0049511317
	Gallons (Brit.)	0.8326747
	Gallons (U.S., dry)	0.85936701
	Gallons (wine)	1
	Gills (U.S.)	32
	Liters	3.785306
	Minims (U.S.)	61,440
	Ounces (U.S., fluid)	128
	Pints (U.S., liq.)	8
	Quarts (U.S., liq.)	4
Gallons (U.S.) of H_2O (4°C.)	Lb. of H_2O	8.34517
Gallons (U.S.) of H_2O (60°F.)	Lb. of H_2O	8.33717
Gallons (U.S.)/day	Cu. ft./hr.	0.0055700231
Gallons (Brit.)/hr.	Cu. meters/min.	7.576812×10^{-5}
Gallons (U.S.)/hr.	Acre-feet/hr.	3.0688833×10^{-6}
	Cu. ft./hr.	0.1336805
	Cu. meters/min.	6.3090197×10^{-5}
	Cu. yd./min.	8.2518861×10^{-5}
	Liters/hr.	3.785306
Gal. (Brit.)/sec.	Cu. cm./sec.	4546.087
Gal. (U.S.)/sec.	Cu. cm./sec.	3785.4118
	Cu. ft./min.	8.020833
	Cu. yd./min.	0.29706790
	Liters/min.	227.1183

To convert from	To	Multiply by
Gammas	Grams	1×10^{-6}
	Micrograms	1
Gausses	E.M. cgs. units of magnetic flux density	1
	E.S. cgs. units	3.335635×10^{-11}
	Gausses (Int.)	0.999670
	Maxwells/sq. cm.	1
	Lines/sq. cm.	1
	Lines/sq. inch	6.4516
Gausses (Int.)	Gausses	1.000330
Gausses/oersted	E.M. cgs. units of permeability	1
	E.S. cgs. units	1.112646×10^{-21}
Geepounds	Slugs	1
	Kilograms	14.5939
Gigameters	Meters	1×10^{9}
Gilberts	Abampere-turns	0.079577472
	Ampere-turns	0.79577472
	E.M. cgs. units of mmf., or magnetic potential	1
	E.S. cgs. units	2.997930×10^{10}
	Gilberts (Int.)	1.000165
Gilberts (Int.)	Gilberts	0.999835
Gilberts/cm.	Ampere-turns/cm.	0.79577472
	Ampere-turns/in.	2.0212678
	Oersteds	1
Gilberts/maxwell	Ampere-turns/weber	7.957747×10^{7}
	E.M. cgs. units of reluctance	1
	E.S. cgs. units	8.987584×10^{20}
Gills (Brit.)	Cu. cm.	142.0652
	Gallons (Brit.)	0.03125
	Gills (U.S.)	1.200949
	Liters	0.1420613
	Ounces (Brit., fluid)	5
	Ounces (U.S., fluid)	4.803764
	Pints (Brit.)	0.25
Gills (U.S.)	Cu. cm.	118.29412
	Cu. inches	7.21875
	Drams (U.S., fluid)	32

To convert from	To	Multiply by
	Gallons (U.S., liq.)	0.03125
	Gills (Brit.)	0.8326747
	Liters	0.1182908
	Minims (U.S.)	1920
	Ounces (U.S., fluid)	4
	Pints (U.S., liq.)	0.25
	Quarts (U.S., liq.)	0.125
Gons (Grades)	Circles	0.0025
	Circumferences	0.0025
	Degrees	0.9
	Minutes	54
	Radians	0.015707963
	Revolutions	0.0025
	Seconds	3240
Grains	Carats (metric)	0.32399455
	Drams (apoth. *or* troy)	0.016666
	Drams (avdp.)	0.036571429
	Dynes	63.5460
	Grams	0.06479891
	Milligrams	64.79891
	Ounces (apoth. *or* troy)	0.0020833
	Ounces (avdp.)	0.0022857143
	Pennyweights	0.041666
	Pounds (apoth. *or* troy)	0.000173611
	Pounds (avdp.)	0.00014285714
	Scruples (apoth.)	0.05
	Tons (metric)	6.479891×10^{-8}
Grains/cu. ft.	Grams/cu. meter	2.2883519
Grains/gal. (U.S.)	Parts/million	17.11854
	Pounds/million gal.	142.8571
Grams-force	Dynes	980.665
	Newtons	9.80665×10^{-3}
Grams	Carats (metric)	5
	Decigrams	10
	Dekagrams	0.1
	Drams (apoth. *or* troy)	0.25720597
	Drams (avdp.)	0.56438339
	Dynes	980.665
	Grains	15.432358
	Kilograms	0.001

To convert from	To	Multiply by
	Micrograms	1×10^{6}
	Myriagrams	0.0001
	Ounces (apoth. *or* troy)	0.032150737
	Ounces (avdp.)	0.035273962
	Pennyweights	0.64301493
	Poundals	0.0709316
	Pounds (apoth. *or* troy)	0.0026792289
	Pounds (avdp.)	0.0022046226
	Scruples (apoth.)	0.77161792
	Tons (metric)	1×10^{-6}
Grams/cm.	Dynes/cm.	980.665
	Grams/inch	2.54
	Kg./km.	100
	Kg./meter	0.1
	Poundals/inch	0.180166
	Pounds/ft.	0.067196898
	Pounds/inch	0.0055997415
	Tons (metric)/km.	0.1
Grams/(cm. × sec.)	Poises	1
	Lb./(ft. × sec.)	0.06719690
Grams/cu. cm.	Dynes/cu. cm.	980.665
	Grains/milliliter	15.43279
	Grams/milliliter	1.000028
	Poundals/cu. inch	1.16236
	Pounds/circ. mil-ft.	3.4049170×10^{-7}
	Pounds/cu. ft.	62.427961
	Pounds/cu. inch	0.036127292
	Pounds/gal. (Brit.)	10.02241
	Pounds/gal. (U.S., dry)	9.7111064
	Pounds/gal. (U.S., liq.)	8.3454044
Grams/cu. meter	Grains/cu. ft.	0.43699572
Grams/liter	Parts/million	1000
	Lb./1000 cu. ft.	0.06242621
	Lb./gal. (U.S.)	8.345171
Grams/milliliter	Grams/cu. cm.	0.999972
	Pounds/cu. ft.	62.42621
	Pounds/gallon (U.S.)	8.345171
Grams/sq. cm.	Atmospheres	0.000967841
	Bars	0.000980665

To convert from	To	Multiply by
	Cm. of Hg (0°C.)	0.0735559
	Dynes/sq. cm.	980.665
	In. of Hg (32°F.)	0.0289590
	Kg./sq. meter	10
	Mm. of Hg (0°C.)	0.735559
	Poundals/sq. inch	0.457623
	Pounds/sq. inch	0.014223343
Grams/ton (long)	Milligrams/kg.	0.98420653
Grams/ton (short)	Milligrams/kg.	1.1023113
Grams-cm.	B.t.u.	9.30113×10^{-8}
	B.t.u. (IST.)	9.29505×10^{-8}
	B.t.u. (mean)	9.28776×10^{-8}
	Cal.	2.34385×10^{-5}
	Cal. (IST.)	2.34231×10^{-5}
	Cal. (mean)	2.34048×10^{-5}
	Cal. (15°C.)	2.34284×10^{-5}
	Cal. (20°C.)	2.34502×10^{-5}
	Kilocal.	2.34385×10^{-8}
	Kilocal. (IST.)	2.34231×10^{-8}
	Kilocal. (mean)	2.34048×10^{-8}
	Dyne-cm.	980.665
	Ergs	980.665
	Foot-poundals	0.00232715
	Foot-pounds	7.2330138×10^{-5}
	Hp.-hours	3.65303×10^{-11}
	Joules	9.80665×10^{-5}
	Kw.-hours	2.72407×10^{-11}
	Kw.-hours (Int.)	2.72362×10^{-11}
	Newton-meters	9.80665×10^{-5}
	Watt-hours	2.72407×10^{-8}
Gram-cm./sec.	B.t.u./sec.	9.30113×10^{-8}
	Cal./sec.	2.34385×10^{-5}
	Ergs-sec.	980.665
	Foot-pounds/sec.	7.2330138×10^{-5}
	Horsepower	1.31509×10^{-7}
	Joules/sec.	9.80665×10^{-5}
	Kilowatts	9.80665×10^{-8}
	Kilowatts (Int.)	9.80503×10^{-8}
	Watts	9.80665×10^{-5}

To convert from	To	Multiply by
Gram/sq. cm.	Pounds/sq. inch	0.000341717
Gram wt.-sec./sq. cm.	Poises	980.665
Gravitational constants	Cm./(sec. × sec.)	980.621
	Ft./(sec. × sec.)	32.1725
Hands	Centimeters	10.16
	Inches	4
Hectares	Acres	2.4710538
	Ares	100
	Sq. cm.	1×10^8
	Sq. feet	107639.10
	Sq. meters	10,000
	Sq. miles	0.0038610216
	Sq. rods	395.36861
Hectograms	Grams	100
	Poundals	7.09316
	Pounds (apoth *or* troy)	0.26792289
	Pounds (avdp.)	0.22046226
Hectoliters	Bushels (Brit.)	2.749694
	Bushels (U.S.)	2.837839
	Cu. cm.	1.00028×10^5
	Cu. feet	3.531566
	Gallons (U.S., liq.)	26.41794
	Liters	100
	Ounces (U.S.) fluid	3381.497
	Pecks (U.S.)	11.35136
Hectometers	Centimeters	10,000
	Decimeters	1000
	Dekameters	10
	Feet	328.08399
	Meters	100
	Rods	19.883878
	Yards	109.3613
Hectowatts	Watts	100
Hefner units	Candles (English)	0.86
	Candles (German)	0.85
	Candles (Int.)	0.90
	10-cp. pentane candles	0.090

To convert from	To	Multiply by
Henries	Abhenries	1×10^9
	E.M. cgs. units	1×10^9
	E.S. cgs. units	1.112646×10^{-12}
	Henries (Int.)	0.999505
	Millihenries	1000
	Mks. (r *or* nr) units	1
	Stathenries	1.112646×10^{-12}
Henries (Int.)	Henries	1.000495
Henries/meter	Cgs. units of permeability	795,774.72
	E.M. cgs. units	795,774.72
	E.S. cgs. units	8.854156×10^{-16}
	Gausses/oersted	795,774.72
	Mks. (nr) units	0.079577472
	Mks. (r) units	1
Hogsheads	Butts (Brit.)	0.5
	Cu. feet	8.421875
	Cu. inches	14,553
	Cu. meters	0.23848094
	Gallons (Brit.)	52.458505
	Gallons (U.S.)	63
	Gallons (wine)	63
	Liters	238.47427
Horsepower	B.t.u. (mean)/hr.	2542.48
	B.t.u./min.	42.4356
	B.t.u. (mean)/sec.	0.706243
	Cal./hr.	6.41616×10^5
	Cal. (IST.)/hr.	6.41196×10^5
	Cal. (mean)/hr.	6.40693×10^5
	Cal./min.	10,693.6
	Cal. (IST.)/min.	10,686.6
	Cal. (mean)/min.	10.678.2
	Ergs/sec.	7.45700×10^9
	Foot-pounds/hr.	1,980,000
	Foot-pounds/min.	33,000
	Foot-pounds/sec.	550
	Horsepower (boiler)	0.0760181
	Horsepower (electric)	0.999598
	Horsepower (metric)	1.01387
	Joules/sec.	745.700
	Kilowatts	0.745700

To convert from	To	Multiply by
	Kilowatts (Int.)	0.745577
	Tons of refrig. (U.S., comm.)	0.21204
	Watts	745.700
Horsepower (boiler)	B.t.u. (mean)/hr.	33,445.7
	Cal./min.	140,671.6
	Cal. (mean)/min.	140,469.4
	Cal. (15°C.)/min.	140,611.1
	Cal. (20°C.)/min.	140,742.2
	Ergs/sec.	9.80950×10^{10}
	Foot-pounds/min.	434,107
	Horsepower	13.1548
	Horsepower (electric)	13.1495
	Horsepower (metric)	13.3372
	Horsepower (water)	13.1487
	Joules/sec.	9809.50
	Kilowatts	9.809.50
	Lb. H_2O evap. per hr. from and at 212°F	34.5
Horsepower (electric)	B.t.u./hr.	2547.16
	B.t.u. (IST.)/hr.	2545.50
	B.t.u. (mean)/hr.	2543.50
	Cal./sec.	178.298
	Kilocal./hr.	641.874
	Ergs/sec.	7.46×10^9
	Foot-pounds/min.	33,013.3
	Foot-pounds/sec.	550.221
	Horsepower	1.00040
	Horsepower (boiler)	0.0760487
	Horsepower (metric)	1.0142777
	Horsepower (water)	0.999942
	Joules/sec.	746
	Kilowatts	0.746
	Watts	746
Horsepower (metric)	B.t.u/hr.	2511.31
	B.t.u. (IST.)/hr.	2509.66
	B.t.u. (mean)/hr.	2507.70
	Cal./hr.	6.32838×10^5

To convert from	To	Multiply by
	Cal. (IST.)/hr.	6.32425×10^5
	Cal. (mean)/hr.	6.31929×10^5
	Ergs/sec.	7.35499×10^9
	Foot-pounds/min.	32,548.6
	Foot-pounds/sec.	542.476
	Horsepower	0.986320
	Horsepower (boiler)	0.0749782
	Horsepower (electric)	0.985923
	Horsepower (water)	0.985866
	Kg.-meter/sec.	75
	Kilowatts	0.735499
	Watts	735.499
Horsepower (water)	Foot-pounds/min.	33,015.2
	Horsepower	1.00046
	Horsepower (boiler)	0.0760531
	Horsepower (electric)	1.00006
	Horsepower (metric)	1.01434
	Kilowatts	0.746043
Horsepower-hours	B.t.u.	2546.14
	B.t.u. (IST.)	2544.47
	B.t.u. (mean)	2542.48
	Cal.	641,616
	Cal. (IST.)	641,196
	Cal. (mean)	640,693
	Foot-pounds	1.98×10^6
	Joules	2.68452×10^6
	Kg.-meters	273,745
	Kw.-hours	0.745700
	Watt-hours	745.700
Hp.-hr./lb.	B.t.u./lb.	2546.14
	Cal./gram	1414.52
	Cu. ft.-(lb./sq. in.)/lb.	13,750
	Foot-pounds/lb.	1,980,000
	Joules/gram	5918.35
Hours (mean solar)	Days (mean solar)	0.0416666
	Days (sidereal)	0.041780746
	Hours (sidereal)	1.00273791
	Minutes (mean solar)	60
	Minutes (sidereal)	60.164275

To convert from	To	Multiply by
	Seconds (mean solar)	3600
	Seconds (sidereal)	3609.8565
	Weeks (mean calendar)	0.0059523809
Hours (sidereal)	Days (mean solar)	0.41552899
	Days (sidereal)	0.0416666
	Hours (mean solar)	0.99726957
	Minutes (mean solar)	59.836174
	Minutes (sidereal)	60
Hundredweights (long)	Kilograms	50.802345
	Pounds	112
	Quarters (Brit., long)	4
	Quarters (U.S., long)	0.2
	Tons (long)	0.05
Hundredweights (short)	Kilograms	45.359237
	Pounds (advp.)	100
	Quarters (Brit., short)	4
	Quarters (U.S., short)	0.2
	Tons (long)	0.044642857
	Tons (metric)	0.045359237
	Tons (short)	0.05
Inches	Ångström units	2.54×10^8
	Centimeters	2.54
	Chains (Gunter's)	0.00126262
	Cubits	0.055555
	Fathoms	0.013888
	Feet	0.083333
	Feet (U.S. Survey)	0.083333167
	Links (Gunter's)	0.126262
	Links (Ramden's)	0.083333
	Meters	0.0254
	Mils	1000
	Picas (printer's)	6.0225
	Points (printer's)	72.27000
	Wavelength of orange-red line of krypton 86	41,929.399

To convert from	To	Multiply by
	Wavelength of the red line of cadmium	39,450.369
	Yards	0.027777
Inches of Hg (32°F.)	Atmospheres	0.0334211
	Bars	0.0338639
	Dynes/sq. cm.	33,863.9
	Ft. of air (1 atm., 60°F.)	926.24
	Ft. of H_2O (39.2°F.)	1.132957
	Grams/sq. cm.	34.5316
	Kg./sq. meter	345.316
	Mm. of Hg (60°C.)	25.4
	Ounces/sq. inch	7.85847
	Pascals	3386.39
Inches of Hg (32°F.)	Pounds/sq. ft.	70.7262
Inches of Hg (60°F.)	Atmospheres	0.0333269
	Dynes/sq. cm.	39,768.5
	Grams/sq. cm.	34.4343
	Mm. of Hg (60°F.)	25.4
	Ounces/sq. inch	7.83633
	Pounds/sq. ft.	70.5269
Inches of H_2O (4°C.)	Atmospheres	0.0024582
	Dynes/sq. cm.	2490.82
	In. of Hg (32°F.)	0.0735539
	Kg./sq. meter	25.3993
	Ounces/sq. ft.	83.2350
	Ounces/sq. inch	0.578020
	Pascals	249.089
	Pounds/sq. ft.	5.20218
	Pounds/sq. inch	0.03612628
Inches/hr.	Cm./hr.	2.54
	Feet/hr.	0.0833333
	Miles/hr.	1.578282×10^{-5}
Inches/min.	Cm./hr.	152.4
	Feet/hr.	5
	Miles/hr.	0.000946969
Joules (abs.)	B.t.u.	0.000948451
	B.t.u. (IST.)	0.000947831

To convert from	To	Multiply by
	B.t.u. (mean)	0.000947088
	Cal.	0.239006
	Cal. (IST.)	0.238849
	Cal. (mean)	0.238662
	Cal. (15°C.)	0.238903
	Cal. (20°C.)	0.239126
	Kilocal. (mean)	0.000238662
	Cu. ft.-atm.	0.000348529
	Ergs	1×10^7
	Foot-poundals	23.730360
	Foot-pounds	0.737562
	Gram-cm.	10,197.16
	Hp.-hours	3.72506×10^{-7}
	Joules (Int.)	0.999835
	Kg.-meters	0.1019716
	Kw.-hours	2.7777×10^{-7}
	Liter-atm.	0.00986895
	Volt-coulombs (Int.)	0.999835
	Watt-hours (abs.)	0.0002777777
	Watt-hours (Int.)	0.000277732
	Watt-sec.	1
	Watt-sec. (Int.)	0.999835
Joules (Int.)	B.t.u.	0.000948608
	B.t.u. (IST.)	0.000947988
	B.t.u. (mean)	0.000947244
	Cal.	0.239045
	Cal. (IST.)	0.238888
	Cal. (mean)	0.238702
	Cu. cm.-atm.	9.87086
	Cu. ft.-atm.	0.000348586
	Dyne-cm.	1.000165×10^7
	Ergs	1.000165×10^7
	Foot-poundals	23.73428
	Foot-pounds	0.737684
	Gram-cm.	10,198.8
	Joules (abs.)	1.000165
	Kw.-hours	2.77824×10^{-7}
	Liter-atm.	0.00987058
	Volt-coulombs	1.000165
	Volt-coulombs (Int.)	1

To convert from	To	Multiply by
	Watt-sec.	1.000165
	Watt-sec. (Int.)	1
Joules/(abcoulomb × °F.)	Joules/(coulomb × °C.)	0.18
Joules/amp.-hr.	Joules/abcoulomb	0.002777
	Joules/statcoulomb	9.265653×10^{-14}
Joules/coulomb	Joules/abcoulomb	10
	Volts	1
Joules/(coulomb × °F.)	Joules/(coulomb × °C.)	1.8
Joules/°C	B.t.u./°F	0.000526917
	Cal./°C.	0.239006
	Cal. (mean)/°C	0.238662
Joules/electronic charge	Joules/abcoulomb	6.24196×10^{19}
Joules/(electronic charge × °C.)	Joules/(coulomb × °C.)	6.24196×10^{18}
Joules/(gram × °C.)	B.t.u./(lb. × °F.)	0.239006
	Cal./(gram × °C.)	0.239006
Joules (Int.)/(gram × °C.)	B.t.u./(lb. × °F.)	0.239045
	Cal. (mean)/(gram × °C.)	0.238702
Joules/sec. (abs.)	B.t.u./min.	0.0569071
	Cal./min	14.3403
	Kilocal./min	0.0143403
	Kilocal. (mean)/min.	0.0143197
	Dyne-cm./sec.	1×10^{7}
	Ergs/sec.	1×10^{7}
	Foot-pounds/sec.	0.737562
	Gram-cm./sec.	10,197.16
	Horsepower	0.00134102
	Watts	1
	Watts (Int.)	0.999835
Joules (Int.)/sec.	B.t.u./min.	0.0569165
	B.t.u. (mean)/min.	0.0568347
	Cal./min.	14.3427
	Kilocal./min.	0.0143427
	Dyne-cm./sec.	1.000165×10^{7}
	Ergs/sec.	1.000165×10^{7}
	Foot-pounds/min.	44.2610

To convert from	To	Multiply by
	Foot-pounds/sec.	0.737684
	Gram-cm./sec.	10,198.8
	Horsepower	0.00134124
	Watts	1.000165
	Watts (Int.)	1
Kilderkins (Brit.)	Cu. cm.	81,829.57
	Cu. feet	2.889784
	Cu. inches	4993.55
	Cu. meters	0.08182957
	Gallons (Brit.)	18
Kilocalories	B.t.u.	3.9683207
	B.t.u. (IST.)	3.96573
	B.t.u. (mean)	3.96262
	B.t.u. (60°F.)	3.96709
	Kilocal.	1000
	Kilocal. (mean)	0.998563
	Kilocal. (15°C.)	0.999570
	Kilocal. (20°C.)	1.00050
	Cu. cm.-atm.	41,292.86
	Ergs	4.184×10^{10}
	Foot-poundals	99,287.8
	Foot-pounds	3085.96
	Gram-cm.	4.26649×10^{7}
	Hp.-hours	0.00155857
	Joules	4184
	Kw.-hours	0.001162222
	Liter-atm.	41.2917
	Watt-hours	1.162222
Kilocalories (mean)	B.t.u.	3.97403
	B.t.u. (IST.)	3.97144
	B.t.u. (mean)	3.9683207
	B.t.u. (60°F.)	3.97280
	Cal.	1001.44
	Cal. (IST.)	1000.78
	Cal. (mean)	1000
	Cal. (15°C.)	1000.10
	Cal. (20°C.)	1001.94
	Ergs	4.19002×10^{10}
	Foot-poundals	99,430.8
	Foot-pounds	3090.40

To convert from	To	Multiply by
	Gram-cm.	4.27263×10^7
	Hp.-hours	0.00156081
	Joules	4190.02
	Kg.-meters	427.263
	Kw.-hours (Int.)	0.00116370
	Liter-atm.	41.3511
	Watt-hours	1.16390
Kilocalories/hr.	Watts	1.162222
Kilocalories/min.	Kg. ice melted/min.	0.012548
	Lb. ice melted/min.	0.027665
	Watts	69.7333
Kilograms-force	Dynes	9.80665×10^5
	Newtons	9.80665
	Pounds-force	2.20462
	Poundals	70.9316
Kilograms-force/sq. cm.	Pascals (N/sq. meter)	98,066.5
Kilograms	Drams (apoth. *or* troy)	257.20597
	Drams (avdp.)	564.38339
	Dynes	980,665
	Grains	15,432.358
	Hundredweights (long)	0.019684131
	Hundredweights (short)	0.022046226
	Ounces (apoth. *or* troy)	32.150737
	Ounces (avdp.)	35.273962
	Pennyweights	643.01493
	Poundals	70.931635
	Pounds (apoth. *or* troy)	2.6792289
	Pounds (avdp.)	2.2046226
	Quarters (Brit., long)	0.078736522
	Quarters (U.S., long)	0.0039368261
	Scruples (apoth.)	771.61792
	Slugs	0.06852177
	Tons (long)	0.00098420653
	Tons (metric)	0.001
	Tons (short)	0.0011023113
Kilograms/cu. meter	Grams/cu. cm.	0.001
	Lb./cu. ft.	0.062427961
	Lb./cu. inch	3.6127292×10^{-5}

To convert from	To	Multiply by
Kg. of ice melted/ hr.	Tons of refrig. (U.S., comm.)	0.026336
Kilograms/sq. cm.	Atmospheres	0.967841
	Bars	0.980665
	Cm. of Hg (0°C.)	73.5559
	Dynes/sq. cm.	980,665
	Ft. of H_2O (39.2°F.)	32.8093
	In. of Hg (32°F.)	28.9590
	Pounds/sq. inch	14.223343
Kilograms/sq. meter	Atmospheres	9.67841×10^{-5}
	Bars	9.80665×10^{-5}
	Dynes/sq. cm.	98.0665
	Ft. of H_2O (39.2°F.)	0.00328093
	Grams/sq. cm.	0.1
	In. of Hg (32°F.)	0.00289590
	Mm. of Hg (0°C.)	0.0735559
	Pounds/sq. ft.	0.20481614
	Pounds/sq. in.	0.0014223343
Kilograms/sq. mm.	Pounds/sq. ft.	204,816.14
	Pounds/sq. in.	1422.3343
	Tons (short)/sq. in.	0.71116716
Kilogram sq. cm.	Pounds sq. ft.	0.0023730360
	Pounds sq. in.	0.34171719
Kilogram-meters	B.t.u. (mean)	0.00928776
	Cal. (mean)	2.34048
	Kilocal. (mean)	0.00234048
	Cu. ft.-atm.	0.00341790
	Dynes-cm.	9.80665×10^{7}
	Ergs	9.80665×10^{7}
	Foot-poundals	232.715
	Foot-pounds	7.23301
	Gram-cm.	100,000
	Hp.-hours	3.65304×10^{-6}
	Joules	9.80665
	Joules (Int.)	9.80503
	Kw.-hours	2.72407×10^{-6}
	Liter-atm.	0.0967814
	Newton-meters	9.80665
	Watt-hours	0.00272407
	Watt-hours (Int.)	0.00272362

To convert from	To	Multiply by
Kilogram-meters/ sec.	Watts	9.80665
Kilolines	Maxwells	1000
	Webers	1×10^{-5}
Kiloliters	Cu. centimeters	1.000028×10^{6}
	Cu. feet	35.31566
	Cu. inches	61,025.45
	Cu. meters	1.000028
	Cu. yards	1.307987
	Gallons (Brit.)	219.9755
	Gallons (U.S., dry)	227.0271
	Gallons (U.S., liq.)	264.1794
	Liters	1000
Kilometers	Astronomical units	6.68878×10^{-9}
	Centimeters	100,000
	Feet	3280.8399
	Feet (U.S. Survey)	3280.833
	Light years	1.05702×10^{-13}
	Meters	1000
	Miles (naut., Int.)	0.53995680
	Miles (statute)	0.62137119
	Myriameters	0.1
	Rods	198.83878
	Yards	1093.6133
Kilometers/hr.	Cm./sec.	27.7777
	Feet/hr.	3280.8399
	Feet/min.	54.680665
	Knots (Int.)	0.53995680
	Meters/sec.	0.277777
	Miles (statute)/hr.	0.62137119
Kilometers/(hr. × sec.)	Cm./(sec. × sec.)	27.7777
	Ft./(sec. × sec.)	0.91134442
	Meters/(sec. × sec.)	0.277777
Kilometers/min.	Cm./sec.	1666.666
	Feet/min.	3280.8399
	Kilometers/hr.	60
	Knots (Int.)	32.397408
	Miles/hr.	37.282272
	Miles/min.	0.62137119

To convert from	To	Multiply by
Kilovolts/cm.	Abvolts/cm.	1×10^{11}
	Microvolts/meter	1×10^{11}
	Millivolts/meter	1×10^{8}
	Statvolts/cm.	3.335635
	Volts/inch	2540
Kilowatts	B.t.u./hr.	3414.43
	B.t.u. (IST.)/hr.	3412.19
	B.t.u. (mean)/hr.	3409.52
	B.t.u. (mean)/min.	56.8253
	B.t.u. (mean)/sec.	0.947088
	Cal. (mean)/hr.	859,184
	Cal. (mean)/min.	14,319.7
	Cal. (mean)/sec.	238.662
	Kilocal. (mean)/hr.	859.184
	Kilocal. (mean)/min.	14.3197
	Kilocal. (mean)/sec.	0.238662
	Cu. ft.-atm./hr.	1254.70
	Ergs/sec.	1×10^{10}
	Foot-poundals/min.	1.42382×10^{6}
	Foot-pounds/hr.	2.65522×10^{6}
	Foot-pounds/min.	44,253.7
	Foot-pounds/sec.	737.562
	Gram-cm./sec.	1.019716×10^{7}
	Horsepower	1.34102
	Horsepower (boiler)	0.101942
	Horsepower (electric)	1.34048
	Horsepower (metric)	1.35962
	Joules/hr.	3.6×10^{6}
	Joules (IST.)/hr.	3.59941×10^{6}
	Joules/sec.	1000
	Kg.-meters/hr.	3.67098×10^{5}
	Kilowatts (Int.)	0.999835
	Watts (Int.)	999.835
Kilowatts (Int.)	B.t.u./hr.	3414.99
	B.t.u. (IST.)/hr.	3412.76
	B.t.u. (mean)/hr.	3410.08
	B.t.u. (mean)/min.	56.8347
	B.t.u. (mean)/sec.	0.947244
	Cal. (mean)/hr.	859,326
	Cal. (mean)/min.	14,322.1

To convert from	To	Multiply by
	Kilocal./hr.	860.563
	Kilocal. (IST.)/hr.	860
	Kilocal (mean)/hr.	859.326
	Cu. cm.-atm./hr.	3.55351×10^7
	Cu. ft.-atm./hr.	1254.91
	Ergs/sec.	1.000165×10^{10}
	Foot-poundals/min.	1.42406×10^6
	Foot-pounds/min.	44,261.0
	Foot-pounds/sec.	737.684
	Gram-cm./sec.	1.01988×10^7
	Horsepower	1.34124
	Horsepower (boiler)	0.101959
	Horsepower (electric)	1.34070
	Horsepower (metric)	1.35985
	Joules/hr.	3.60059×10^6
	Joules (Int.)/hr.	3.6×10^6
	Kg.-meters/hr.	367,158
	Kilowatts	1.000165
Kilowatt-hours	B.t.u. (mean)	3409.52
	Cal. (mean)	859,184
	Foot-pounds	2.65522×10^6
	Hp.-hours	1.34102
	Joules	3.6×10^6
	Kg.-meters	367,098
	Lb. H_2O evap. from and at 212°F	3.5168
	Watt-hours	1000
	Watt-hours (Int.)	999.835
Kilowatt-hours (Int.)	B.t.u. (mean)	3410.08
	Cal. (IST.)	860,000
	Cal. (mean)	859,326
	Cu., cm.-atm.	3.55351×10^7
	Cu. ft.-atm.	1254.91
	Foot-pounds	2.65566×10^6
	Hp.-hours	1.34124
	Joules	3.60059×10^6
	Joules (Int.)	3.6×10^6
	Kg.-meters	367,158

To convert from	To	Multiply by
Kw.-hr./gram	B.t.u./lb.	1.54876×10^6
	B.t.u. (IST.)/lb.	1.54774×10^6
	B.t.u. (mean)/lb.	1.54653×10^6
	Cal./gram	860,421
	Cal. (mean)/gram	859,184
	Cu. cm.-atm./gram	3.55292×10^7
	Cu. ft.-atm./lb.	569,124
	Hp.-hr./lb.	608.277
	Joules/gram	3.6×10^6
Knots (Int.)	Cm./sec.	51.4444
	Feet/hr.	6076.1155
	Feet/min.	101.26859
	Feet/sec.	1.6878099
	Kilometers/hr.	1.852
	Meters/min.	30.8666
	Meters/sec.	0.514444
	Miles (naut., Int.)/hr.	1
	Miles (statute)/hr.	1.1507794
Lamberts	Candles/sq. cm.	0.31830989
	Candles/sq. ft.	295.71956
	Candles/sq. inch	2.0536081
	Foot-lamberts	929.0304
	Lumens/sq. cm.	1
Lasts (Brit.)	Liters	2909.414
Leagues (naut., Brit.)	Feet	18,240
	Kilometers	5.559552
	Leagues (naut., Int.)	1.0006393
	Leagues (statute)	1.151515
	Miles (statute)	3.454545
Leagues (naut., Int.)	Fathoms	3038.0577
	Feet	18,228.346
	Kilometers	5.556
	Leagues (statute)	1.1507794
	Miles (statute)	3.4523383
Leagues (statute)	Fathoms	2640
	Feet	15,840
	Kilometers	4.828032
	Leagues (naut., Int.)	0.86897625

To convert from	To	Multiply by
	Miles (naut., Int.)	2.6069287
	Miles (statute)	3
Light years	Astronomical units	63,279.5
	Kilometers	9.46055×10^{12}
	Miles (statute)	5.87851×10^{12}
Lines	Maxwells	1
Lines (Brit.)	Centimeters	0.211666
	Inches	0.083333
Lines/sq. cm.	Gausses	1
Lines/sq. inch	Gausses	0.15500031
	Webers/sq. inch	1×10^{-8}
Links (Gunter's)	Chains (Gunter's)	0.01
	Feet	0.66
	Feet (U.S. Survey)	0.65999868
	Inches	7.92
	Meters	0.201168
	Miles (statute)	0.000125
	Rods	0.04
Links (Ramden's)	Centimeters	30.48
	Chains (Ramden's)	0.01
	Feet	1
	Inches	12
Liters	Bushels (Brit.)	0.2749694
	Bushels (U.S.)	0.02837839
	Cu. centimeters	1000.028
	Cu. feet	0.03531566
	Cu. inches	61.02545
	Cu. meters	0.001000028
	Cu. yards	0.001307987
	Drams (U.S., fluid)	270.5198
	Gallons (Brit.)	0.2199755
	Gallons (U.S., dry)	0.2270271
	Gallons (U.S., liq.)	0.2641794
	Gills (Brit.)	7.039217
	Gills (U.S.)	8.453742
	Hogsheads	0.004193325
	Minims (U.S.)	16,231.19
	Ounces (Brit., fluid)	35.19609
	Ounces (U.S., fluid)	33.81497
	Pecks (Brit.)	0.1099878

To convert from	To	Multiply by
	Pecks (U.S.)	0.1135136
	Pints (Brit.)	1.759804
	Pints (U.S., dry)	1.816217
	Pints (U.S., liq.)	2.113436
	Quarts (Brit.)	0.8799021
	Quarts (U.S., dry)	0.9081084
	Quarts (U.S., liq.)	1.056718
Liters/min.	Cu. ft./min.	0.03531566
	Cu. ft./sec.	0.0005885943
	Gal. (U.S., liq.)/min.	0.2641794
Liters/sec.	Cu. ft./min.	2.118939
	Cu. ft./sec.	0.03531566
	Cu. yards/min.	0.07847923
	Gal. (U.S., liq.)/min.	15.85077
	Gal. (U.S., liq.)/sec.	0.2641794
Liter-atm.	B.t.u.	0.0961045
	B.t.u. (IST.)	0.0960417
	B.t.u. (mean)	0.0959664
	Cal.	24.2179
	Cal. (IST.)	24.2021
	Cal. (mean)	24.1831
	Cu. ft.-atm.	0.0353157
	Foot-poundals	2404.55
	Foot-pounds	74.7356
	Hp.-hours	3.77452×10^{-5}
	Joules	101.328
	Joules (Int.)	101.311
	Kg.-meters	10.3326
	Kw.-hours	2.81466×10^{-5}
Liter-atm. (lat. 45°)	Joules	101.323
Lumens	Candle power (spher.)	0.079577472
Lumens (at 5550 Å)	Watts	0.0014705882
Lumens/sq. cm.	Lamberts	1
	Phots	1
Lumens/(sq. cm. × steradian)	Lamberts	3.1415927
Lumens/sq. ft.	Foot-candles	1
	Foot-lamberts	1
	Lumens/sq. meter	10.763910

To convert from	To	Multiply by
Lumens/(sq. ft. × steradian)	Millilamberts	3.3815822
Lumens/sq. meter	Foot-candles	0.09290304
	Lumens/sq. ft.	0.09290304
	Phots	0.0001
Lux	Foot-candles	0.09290304
	Lumens/sq. meter	1
	Phots	0.0001
Maxwells	E.M. cgs. units of induction flux	1
	E.S. cgs. units	3.335635×10^{-11}
	Gauss-sq. cm.	1
	Lines	1
	Maxwells (Int.)	0.999670
	Volt-seconds	1×10^{-8}
	Webers	1×10^{-8}
Maxwells (Int.)	Maxwells	1.000330
Maxwells/sq. cm.	Maxwells/sq. in.	6.4516
	Maxwells (Int.)/sq. cm.	0.999670
Maxwells (Int.)/sq. cm.	Maxwells/sq. cm.	1.000330
Maxwells/sq. inch	Maxwells/sq. cm.	0.15500031
Megalines	Maxwells	1×10^{6}
MegaPascals	Bars	10
	Newtons/sq. mm.	1
	Pascals	1×10^{6}
Megmhos/cm.	Abmhos/cm.	0.001
	Megmhos/inch cube	2.54
	$(\text{Microhm-cm.})^{-1}$	1
Megmhos/inch	Megmhos/cm.	0.39370079
	$(\text{Microhm-inches})^{-1}$	1
Megohms	Microhms	1×10^{12}
	Ohms	1×10^{6}
	Statohms	1.112646×10^{-6}
Megohms^{-1}	Micromhos	1
Meters	Ångström units	1×10^{10}
	Centimeters	100
	Chains (Gunter's)	0.049709695

To convert from	To	Multiply by
	Chains (Ramden's)	0.032808399
	Fathoms	0.54680665
	Feet	3.2808399
	Feet (U.S. Survey)	3.280833
	Furlongs	0.0049709695
	Inches	39.370079
	Kilometers	0.001
	Links (Gunter's)	4.9709695
	Links (Ramden's)	3.2808399
	Megameters	1×10^{-6}
	Miles (naut., Brit.)	0.00053961182
	Miles (naut., Int.)	0.00053995680
	Miles (statute)	0.00062137119
	Millimeters	1000
	Millimicrons	1×10^{9}
	Mils	39,370.079
	Rods	0.19883878
	Yards	1.0936133
Meters of Hg (0°C.)	Atmospheres	1.3157895
	Ft. of H_2O (60°F.)	44.6474
	In. of Hg (32°F.)	39.370079
	Kg./sq. cm.	1.35951
	Pounds/sq. inch	19.3368
Meters/hr.	Feet/hr.	3.2808399
	Feet/min.	0.054680665
	Knots (Int.)	0.00053995680
	Miles (statute)/hr.	0.00062137119
Meters/min.	Cm./sec.	1.666666
	Feet/min.	3.2808399
	Feet/sec.	0.054680665
	Kilometers/hr.	0.06
	Knots (Int.)	0.032397408
	Miles (statute)/hr.	0.037282272
Meters/sec.	Feet/min.	196.85039
	Feet/sec.	3.2808399
	Kilometers/hr.	3.6
	Kilometers/min.	0.06
	Miles (statute)/hr.	2.2369363
Meters/(sec. × sec.)	Kilometers/(hr. × sec.)	3.6
	Miles/(hr. × sec.)	2.2369363

To convert from	To	Multiply by
Meter-candles	Lumens/sq. meter	1
Mhos	Abmhos	1×10^{-9}
	Cgs. units of conductance	1
	E.M. cgs. units	1×10^{-9}
	E.S. cgs. units	8.987584×10^{11}
	Mhos (Int.)	1.000495
	Mks. (r *or* nr) units	1
	$Ohms^{-1}$	1
	Siemen's units	1
	Statmhos	8.987584×10^{11}
Mhos (Int.)	Abmhos	9.99505×10^{-10}
	Mhos	0.999505
Mhos/meter	Abmhos/cm.	1×10^{-11}
	Mhos (Int.)/meter	1.000495
Mho-ft./circ. mil	Mhos/cm.	6.0153049×10^{6}
Microfarads	Abfarads	1×10^{-15}
	Farads	1×10^{-6}
	Statfarads	8.987584×10^{5}
Micrograms	Grams	1×10^{-6}
	Milligrams	0.001
Microhenries	Henries	1×10^{-6}
	Stathenries	1.112646×10^{-18}
Microhms	Abohms	1000
	Megohms	1×10^{-12}
	Ohms	1×10^{-6}
	Statohms	1.112646×10^{-18}
Microhm-cm.	Abohm-cm.	1000
	Circ. mil-ohms/ft.	6.0153049
	Microhm-inches	0.39370079
	Ohm-cm.	1×10^{-6}
Microhm-inches	Circ. mil-ohms/ft.	15.278875
	Michrom-cm.	2.54
Micromicrofarads	Farads	1×10^{-12}
Micromicrons	Ångström units	0.01
	Centimeters	1×10^{-10}
	Inches	$3.9370079 \times 10^{-11}$
	Meters	1×10^{-12}
	Microns	1×10^{-6}
Microns	Ångström units	10,000

To convert from	To	Multiply by
Microns	Centimeters	0.0001
	Feet	3.2808399×10^{-5}
	Inches	3.9370079×10^{-5}
	Meters	1×10^{-6}
	Millimeters	0.001
	Millimicrons	1000
Miles (naut., Brit.)	Cable lengths (Brit.)	8.4444
	Fathoms	1013.333
	Feet	6080
	Meters	1853.184
	Miles (Adm., Brit.)	1
	Miles (naut., Int.)	1.0006393
	Miles (statute)	1.151515
Miles (naut., Int.)	Cable lengths	8.4390493
	Fathoms	1.012.6859
	Feet	6076.1155
	Feet (U.S. Survey)	6076.1033
	Kilometers	1.852
	Leagues (naut., Int.)	0.333333
	Meters	1852
	Miles (geographical)	1
	Miles (naut., Brit.)	0.99936110
	Miles (statute)	1.1507794
Miles (statute)	Centimeters	160,934.4
	Chains (Gunter's)	80
	Chains (Ramden's)	52.8
	Feet	5280
	Feet (U.S. Survey)	5279.9894
	Furlongs	8
	Inches	63,360
	Kilometers	1.609344
	Light years	1.70111×10^{-12}
	Links (Gunter's)	8000
	Meters	1609.344
	Miles (naut., Brit.)	086842105
	Miles (naut., Int.)	0.86897624
	Myriameters	0.1609344
	Rods	320
	Yards	1760

To convert from	To	Multiply by
Miles/hr.	Cm./sec.	44.704
	Feet/hr.	5280
	Feet/min.	88
	Feet/sec.	1.466666
	Kilometers/hr.	1.609344
	Knots (Int.)	0.86897624
	Meters/min.	26.8224
	Miles/min.	0.0166666
Miles/(hr. × min.)	Cm./(sec. × sec.)	0.7450666
Miles/(hr. × sec.)	Cm./(sec. × sec.)	44.704
	Ft./(sec. × sec.)	1.466666
	Kilometers/(hr. × sec.)	1.609344
	Meters/(sec. × sec.)	0.44704
Miles/min.	Cm./sec.	2682.24
	Feet/hr.	316,800
	Feet/sec.	88
	Kilometers/min.	1.609344
	Knots (Int.)	52.138574
	Meters/min.	1609.344
	Miles/hr.	60
Millibars	Atmospheres	0.000986923
	Bars	0.001
	Baryes	1000
	Dynes/sq. cm.	1000
	Grams/sq. cm.	1.019716
	In. of Hg (32°F.)	0.0295300
	Pascals	100
	Pounds/sq. ft.	2.088543
	Pounds/sq. inch	0.0145038
Milligrams	Carats (1877)	0.004871
	Carats (metric)	0.005
	Drams (apoth. *or* troy)	0.00025720597
	Drams (advp.)	0.00056438339
	Grains	0.015432358
	Grams	0.001
	Ounces (apoth. *or* troy)	3.2150737×10^{-5}
	Ounces (avdp.)	3.5273962×10^{-5}
	Pennyweights	0.00064301493
	Pounds (apoth. *or* troy)	2.6792289×10^{-6}

To convert from	To	Multiply by
	Pounds (avdp.)	2.2046226×10^{-6}
	Scruples (apoth.)	0.00077161792
Milligrams/assay ton	Milligrams/kg.	34.285714
	Ounces (troy)/ton (avdp.)	1
Milligrams/gm.	Dynes/cm.	0.980665
	Pounds/inch	5.5997415×10^{-6}
Milligrams/gram	Carats (parts gold per 24 of mixture)	0.024
	Grams/ton (short)	907.18474
	Milligrams/assay ton	29.166666
	Ounces (avdp.)/ton (long)	35.84
	Ounces (avdp.)/ton (short)	32
	Ounces (troy)/ton (long)	32.6666
	Ounces (troy)/ton (short)	29.1666
Milligrams/inch	Dynes/cm.	0.386089
	Dynes/inch	0.980665
	Grams/cm.	0.00039370079
	Grams/inch	0.0001
Milligrams/kg.	Pounds (avdp.)/ton (short)	0.002
Milligrams/liter	Grains/gal. (U.S.)	0.05841620
	Grams/liter	0.001
	Parts/million	1
	Lb./cu. ft.	6.242621×10^{-5}
Milligrams/mm.	Dynes/cm.	9.80665
Millihenries	Abhenries	1×10^{6}
	Henries	0.001
	Stathenries	1.112646×10^{-15}
Millilamberts	Candles/sq. cm.	0.00031830989
	Candles/sq. inch	0.0020536081
	Foot-lamberts	0.9290304
	Lamberts	0.001
	Lumens/sq. cm.	0.001
	Lumens/sq. ft.	0.9290304
Milliliters	Cu. cm.	1.000028
	Cu. inches	0.06102545
	Drams (U.S., fluid)	0.2705198
	Gills (U.S.)	0.008453742
	Liters	0.001
	Minims (U.S.)	16.23119
	Ounces (Brit., fluid)	0.03519609

To convert from	To	Multiply by
	Ounces (U.S., fluid)	0.03381497
	Pints (Brit.)	0.001759804
	Pints (U.S., liq.)	0.002113436
Millimeters	Ångström units	1×10^{7}
	Centimeters	0.1
	Decimeters	0.01
	Dekameters	0.0001
	Feet	0.0032808399
	Inches	0.039370079
	Meters	0.001
	Microns	1000
	Mils	39.370079
	Wavelength of orange-red line of krypton 86	1650.76373
	Wavelength of red line of cadmium	1553.16413
Millimeters of Hg (0°C)	Atmospheres	0.0013157895
	Bars	0.00133322
	Dynes/sq. cm.	1333.224
	Grams/sq. cm.	1.35951
	Kg./sq. meter	13.5951
	Pascals	133.3224
	Pounds/sq. ft.	2.78450
	Pounds/sq. inch	0.0193368
	Torrs	1
Millimeters of H_2O (4°C)	Pascals	9.80665
Millimicrons	Ångström units	10
	Centimeters	1×10^{-7}
	Inches	3.9370079×10^{-8}
	Microns	0.001
	Millimeters	1×10^{-6}
	Nanometers	1
Milliphots	Foot-candles	0.9290304
	Lumens/sq. ft.	0.9290304
	Lumens/sq. meter	10
	Lux	10
	Phots	0.001

To convert from	To	Multiply by
Millivolts	Statvolts	3.335635×10^{-6}
	Volts	0.001
Minims (Brit.)	Cu. cm.	0.05919385
	Cu. inches	0.003612230
	Milliliters	0.5919219
	Ounces (Brit., fluid)	0.0020833333
	Scruples (Brit., fluid)	0.05
Minims (U.S.)	Cu. cm.	0.061611520
	Cu. inches	0.0037597656
	Drams (U.S., fluid)	0.0166666
	Gallons (U.S., liq.)	1.6276042×10^{-5}
	Gills (U.S.)	0.0005208333
	Liters	6.160979×10^{-5}
	Milliliters	0.06160979
	Ounces (U.S., fluid)	0.002083333
	Pints (U.S., liq.)	0.0001302083
Minutes (angular)	Degrees	0.0166666
	Quadrants	0.000185185
	Radians	0.00029088821
	Seconds (angular)	60
Minutes (mean solar)	Days (mean solar)	0.0006944444
	Days (sidereal)	0.00069634577
	Hours (mean solar)	0.0166666
	Hours (sidereal)	0.016712298
	Minutes (sidereal)	1.00273791
Minutes (sidereal)	Days (mean solar)	0.00069254831
	Minutes (mean solar)	0.99726957
	Months (mean calendar)	2.2768712×10^{-5}
	Seconds (sidereal)	60
Minutes/cm.	Radians/cm.	0.00029088821
Months (lunar)	Days (mean solar)	29.530588
	Hours (mean solar)	708.73411
	Minutes (mean solar)	42524.047
	Seconds (mean solar)	2.5514428×10^{6}
	Weeks (mean calendar)	4.2186554
Months (mean calendar)	Days (mean solar)	30.416666
	Hours (mean solar)	730
	Months (lunar)	1.0300055

To convert from	To	Multiply by
	Weeks (mean calendar)	4.3452381
	Years (calendar)	0.08333333
	Years (sidereal)	0.083274845
	Years (tropical)	0.083278075
Myriagrams	Grams	10,000
	Kilograms	10
	Pounds (avdp.)	22.046226
Nanometers	Ångström	10
	Micrometer	0.001
	Mil	3.937008×10^{-5}
	Millimicron	1
Newtons	Dynes	1×10^{5}
	Kilograms-force	0.1019716
	Poundals	7.23301
	Pounds-force	0.224809
Newton-meters	Dyne-cm.	1×10^{7}
	Gram-cm.	10,197.162
	Kg.-meters	0.10197162
	Pound-feet	0.73756215
Newtons/sq. meter	Pascals	1
Newtons/sq. mm	MegaPascals	1
Noggins (Brit.)	Cu. cm.	142.0652
	Gallons (Brit.)	0.03125
	Gills (Brit.)	1
Oersteds	Ampere-turns/inch	2.0212678
	Ampere-turns/meter	79.577472
	E.M. cgs. units of magnetic field intensity	1
	E.S. cgs. units	2.997930×10^{10}
	Gilberts/cm.	1
	Oersteds (Int.)	1.000165
Oersteds (Int.)	Oersteds	0.999835
Ohms	Abohms	1×10^{9}
	Cgs. units of resistance	1
	Megohms	1×10^{-6}
	Microhms	1×10^{6}
	Ohms (Int.)	0.999505
	Statohms	1.112646×10^{-12}

To convert from	To	Multiply by
Ohms (Int.)	Ohms	1.000495
Ohms (mil, foot)	Circ. mil-ohms/ft.	1
	Ohm-cm.	1.6624261×10^{-7}
Ohm-cm.	Circ. mil-ohms/ft.	6.0153049×10^{6}
	Microhm-cm.	1×10^{6}
	Ohm-inches	0.39370079
Ohm-inches	Ohm-cm.	2.54
Ohm-meters	Abohm-cm.	1×10^{11}
	E.M. cgs. units	1×10^{11}
	E.S. cgs. units	1.112646×10^{-10}
	Mks. units	1
	Statohm-cm.	1.112646×10^{-10}
Ounces (apoth. *or* troy)	Dekagrams	1.7554286
	Drams (apoth. *or* troy)	8
Drams	(avdp.)	17.554286
	Grains	480
	Grams	31.103486
	Milligrams	31,103.486
	Ounces (avdp.)	1.0971429
	Pennyweights	20
	Pounds (apoth. *or* troy)	0.0833333
	Pounds (avdp.)	0.068571429
	Scruples (apoth.)	24
	Tons (short)	3.4285714×10^{-5}
Ounces (advp.)	Drams (apoth. *or* troy)	7.291666
	Drams (avdp.)	16
	Grains	437.5
	Grams	28.349523
	Hundredweights (long)	0.00055803571
	Hundredweights (short)	0.000625
	Ounces (apoth. *or* troy)	0.9114583
	Pennyweights	18.229166
	Pounds (apoth. *or* troy)	0.075954861
	Pounds (avdp.)	0.0625
	Scruples (apoth.)	21.875
	Tons (long)	2.7901786×10^{-5}
	Tons (metric)	2.8349527×10^{-5}
	Tons (short)	3.125×10^{-5}

To convert from	To	Multiply by
Ounces (Brit., fluid)	Cu. cm.	28.41305
	Cu. inches	1.733870
	Drachms (Brit., fluid)	8
	Drams (U.S., fluid)	7.686075
	Gallons (Brit.)	0.00625
	Milliliters	28.41225
	Minims (Brit.)	480
	Ounces (U.S., fluid)	0.9607594
Ounces (U.S., fluid)	Cu. cm.	29.573730
	Cu. inches	1.8046875
	Cu. meters	2.9573730×10^{-5}
	Drams (U.S., fluid)	8
	Gallons (U.S., dry)	0.0067138047
	Gallons (U.S., liq.)	0.0078125
	Gills (U.S.)	0.25
	Liters	0.029572702
	Minims (U.S.)	480
	Ounces (Brit., fluid)	1.040843
	Pints (U.S., liq.)	0.625
	Quarts (U.S., liq.)	0.03125
Ounces/sq. inch	Dynes/sq. cm.	4309.22
	Grams/sq. cm.	4.3941849
	In. of H_2O (39.2°F.)	1.73004
	In. of H_2O (60°F.)	1.73166
	Pounds/sq. ft.	9
	Pounds/sq. inch.	0.0625
Ounces (avdp.)/ton (long)	Milligrams/kg.	27.901786
Ounces (avdp.)/ton (short)	Milligrams/kg.	31.25
Paces	Centimeters	76.2
	Chains (Gunter's)	0.0378788
	Chains (Ramden's)	0.025
	Feet	2.5
	Hands	7.5
	Inches	30
	Ropes (Brit.)	0.125

To convert from	To	Multiply by
Palms	Centimeters	7.62
	Chains (Ramden's)	0.0025
	Cubits	0.1666666
	Feet	0.25
	Hands	0.75
	Inches	3
Parsecs	Kilometers	3.08374×10^{12}
	Light years	3.26164
	Miles (statute)	1.91615×10^{12}
Parts/million	Grains/gal. (Brit.)	0.07015488
	Grains/gal. (U.S.)	0.05841620
	Grams/liter	0.001
	Milligrams/liter	1
Pascals	Atmospheres	9.869233×10^{-6}
	Bars	1×10^{-5}
	Dyne/sq. cm.	10
	Feet of H_2O (conv.)	3.34552×10^{-4}
	Inches of Hg (conv.)	2.95300×10^{-4}
	Inches of H_2O (conv.)	4.01463×10^{-3}
	Kilograms-force/sq. cm.	0.01972×10^{-5}
	MegaPascals	1×10^{-6}
	Millibars	0.01
	Mm. of Hg (conv.)	7.50062×10^{-3}
	Mm. of H_2O (conv.)	0.101972
	Newtons/sq.-meter	1
	Newtons/sq. mm.	1×10^{-6}
	Poundals/sq. ft.	0.671969
	Pounds-force/sq. ft.	0.0208854
	Pounds-force/sq. inch	1.45038×10^{-4}
	Tons	7.50062×10^{-3}
Pecks (Brit.)	Bushels (Brit.)	0.25
	Coombs (Brit.)	0.0625
	Cu. cm.	9092.175
	Cu. inches	554.8385
	Gallons (Brit.)	2
	Gills (Brit.)	64
	Hogsheads	0.03812537
	Kilderkins (Brit.)	0.111111
	Liters	9.091920
	Pints (Brit.)	16

To convert from	To	Multiply by
	Quarterns (Brit., dry)	4
	Quarters (Brit., dry)	0.03125
	Quarts (Brit.)	8
	Quarts (U.S., dry)	8.256449
Pecks (U.S.)	Barrels (U.S., dry)	0.076191185
	Bushels (U.S.)	0.25
	Cu. cm.	8809.7675
	Cu. feet	0.311114005
	Cu. inches	537.605
	Gallons (U.S., dry)	2
	Gallons (U.S., liq.)	2.3272944
	Liters	8.809521
	Pints (U.S., dry)	16
	Quarts (U.S., dry)	8
Pennyweights	Drams (apoth. *or* troy)	0.4
	Drams (avdp.)	0.87771429
	Grains	24
	Grams	1.55517384
	Ounces (apoth. *or* troy)	0.05
	Ounces (avdp.)	0.054857143
	Pounds (apoth. *or* troy)	0.0041666
	Pounds (avdp.)	0.0034285714
Perches (masonry)	Cu. feet	24.75
Phots	Foot-candles	929.0304
	Lumens/sq. cm.	1
	Lumens/sq. meter	10,000
	Lux	10,000
Picas (printer's)	Centimeters	0.42175170
	Inches	0.166044
Pints (Brit.)	Cu. cm.	568.26092
	Gallons (Brit.)	0.125
	Gills (Brit.)	4
	Gills (U.S.)	4.803797
	Liters	0.5682450
	Minims (Brit.)	9600
	Ounces (Brit., fluid)	20
	Pints (U.S., dry)	1.032056
	Pints (U.S., liq.)	1.200949
	Quarts (Brit.)	0.5
	Scruples (Brit., fluid)	480

To convert from	To	Multiply by
Pints (U.S., dry)	Bushels (U.S.)	0.015625
	Cu. cm.	550.61047
	Cu. inches	33.6003125
	Gallons (U.S., dry)	0.125
	Gallons (U.S., liq.)	0.14545590
	Liters	0.5505951
	Pecks (U.S.)	0.0625
	Quarts (U.S., dry)	0.5
Pints (U.S., liq.)	Cu. cm.	473.17647
	Cu. feet	0.016710069
	Cu. inches	28.875
	Cu. yards	0.00061889146
	Drams (U.S., fluid)	128
	Gallons (U.S., liq.)	0.125
	Gills (U.S.)	4
	Liters	0.4731632
	Milliliters	473.1632
	Minims (U.S.)	7680
	Ounces (U.S., fluid)	16
	Pints (Brit.)	0.8326747
	Quarts (U.S., liq.)	0.5
Planck's constant	Erg-seconds	6.6255×10^{-27}
	Joule-seconds	6.6255×10^{-34}
	Joule-sec./Avog. No. (chem.)	3.9905×10^{-10}
Points (printer's)	Centimeters	0.03514598
	Inches	0.013837
	Picas	0.0833333
Poises	Cgs. units of absolute viscosity	1
	Grams/(cm. × sec.)	1
Poise-cu. cm./gram	Sq. cm./sec	1
Poise-cu. ft./lb.	Sq. cm./sec.	62.427960
Poise-cu. in./gram	Sq. cm./sec	16.387064
Poles/sq. cm.	E.M. cgs. units of magnetization	1
Pottles (Brit.)	Gallons (Brit.)	0.5
	Liters	2.272980

To convert from	To	Multiply by
Poundals	Grams-force	14.0981
	Newtons	0.1382550
	Pounds-force	0.0310810
Poundals/sq. ft.	Pascals	1.488164
Pounds (apoth. *or* troy)	Drams (apoth. *or* troy)	96
	Drams (avdp.)	210.65143
	Grains	5760
	Grams	373.24172
	Kilograms	0.37324172
	Ounces (apoth. *or* troy)	12
	Ounces (avdp.)	13.165714
	Pennyweights	240
	Pounds (avdp.)	0.8228571
	Scruples (apoth.)	288
	Tons (long)	0.00036734694
	Tons (metric)	0.00037324172
	Tons (short)	0.00041142857
Pounds (avdp.)	Drams (apoth. *or* troy)	116.6666
	Drams (avdp.)	256
	Grains	7000
	Grams	453.59237
	Hundredweights (long)	0.00892857
	Hundredweights (short)	0.01
	Kilograms	0.45359237
	Ounces (apoth. *or* troy)	14.583333
	Ounces (avdp.)	16
	Pound-force	1
	Pennyweights	291.6666
	Poundals	32.1740
	Pounds (apoth. *or* troy)	1.215277
	Scruples (apoth.)	350
	Slugs	0.0310810
	Tons (long)	0.00044642857
	Tons (metric)	0.00045359237
	Tons (short)	0.0005
Pounds-force	Kilograms-force	0.453592
	Newtons	4.44822

To convert from	To	Multiply by
	Poundals	32.1740
	Pounds (avdp.)	1
Pounds-force/sq. ft.	Pascals	47.8803
Pounds-force/sq. inch	Pascals	6894.76
	MegaPascals	0.00689476
Pounds of H_2O evap. from and at 212°F.	B.t.u.	970.9
	B.t.u. (IST.)	970.2
	B.t.u. (mean)	969.4
	Joules	1.0237×10^6
	Joules (Int.)	1.0234×10^6
Pounds/cu. ft.	Grams/cu. cm.	0.016018463
	Kg./cu. meter	16.018463
Pounds/cu. inch	Grams/cu. cm.	27.679905
	Grams/liter	27.68068
	Kg./cu. meter	27679.905
Pounds/gal. (Brit.)	Pounds/cu. ft.	6.228839
Pounds/gal. (U.S., liq.)	Grams/cu. cm.	0.11982643
	Pounds/cu. ft.	7.4805195
Pounds/inch	Grams/cm.	178.57967
	Grams/ft.	5443.1084
	Grams/inch	453.59237
	Ounces/cm.	6.2992
	Ounces/inch	16
	Pounds/meter	39.370079
Pounds/minute	Kilograms/hr.	27.2155422
	Kilograms/min.	0.45359237
Pounds of H_2O (39.2°F.)/minute	Cu. ft./min.	0.01601891
	Gal. (U.S.)/min.	0.1198298
	Liters/min.	0.45359237
Pounds/sq. ft.	Atmospheres	0.000472541
	Bars	0.000478803
	Cm. of Hg (0°C.)	0.0359131
	Dynes/sq. cm.	478.803
	Ft. of air (1 atm., 60°F.)	13.096
	Grams/sq. cm.	0.48824276

To convert from	To	Multiply by
	In. of Hg (32°F.)	0.0141390
	In. of H_2O (39.2°F.)	0.192227
	Kg./sq. meter	4.8824276
	Mm. of Hg (0°C.)	0.359131
Pounds/sq. inch	Atmospheres	0.0680460
	Bars	0.0689476
	Cm. of Hg (0°C.)	5.17149
	Cm. of H_2O (4°C.)	70.3089
	Dynes/sq. cm.	68,947.6
	Grams/sq. cm.	70.306958
	In. of Hg (32°F.)	2.03602
	In. of H_2O (39.2°F.)	27.6807
	Kg./sq. cm.	0.070306958
	Mm. of Hg (0°C.)	51.7149
Pounds-force-sec/sq. ft.	Poises	478.803
Pounds-force-sec/sq. in.	Poises	68,947.6
Puncheons (Brit.)	Cu. meters	0.31797510
	Gallons (Brit.)	69.94467
	Gallons (U.S.)	84
Quadrants	Minutes	5400
	Radians	1.5707963
Quarterns (Brit., dry)	Buckets (Brit.)	0.125
	Bushels (Brit.)	0.0625
	Cu. cm.	2273.044
	Gallons (Brit.)	0.5
	Liters	2.272980
	Pecks (Brit.)	0.25
Quartens (Brit., liq.)	Cu. cm.	142.0652
	Gallons (Brit.)	0.03125
	Liters	0.1420613
Quarters (U.S., long)	Kilograms	254.0117272
	Pounds (avdp.)	560
Quarters (U.S., short)	Kilograms	226.796185
	Pounds	500

To convert from	To	Multiply by
Quarts (Brit.)	Cu. cm.	1136.522
	Cu. inches	69.35482
	Gallons (Brit.)	0.25
	Gallons (U.S., liq.)	0.3002373
	Liters	1.136490
	Quarts (U.S., dry)	1.032056
	Quarts (U.S., liq.)	1.200949
Quarts (U.S., dry)	Bushels (U.S.)	0.03125
	Cu. cm.	1101.2209
	Cu. feet	0.038889251
	Cu. inches	67.200625
	Gallons (U.S., dry)	0.25
	Gallons (U.S., liq.)	0.29091180
	Liters	1.1011901
	Pecks (U.S.)	0.125
	Pints (U.S., dry)	2
Quarts (U.S., liq.)	Cu. cm.	946.35295
	Cu. feet	0.033420136
	Cu. inches	57.75
	Drams (U.S., fluid)	256
	Gallons (U.S., dry)	0.21484175
	Gallons (U.S., liq.)	0.25
	Gills (U.S.)	8
	Liters	0.9463264
	Ounces (U.S., fluid)	32
	Pints (U.S., liq.)	2
	Quarts (Brit.)	0.8326747
	Quarts (U.S., dry)	0.8593670
Quintals (metric)	Grams	100,000
	Hundredweights (long)	1.9684131
	Kilograms	100
	Pounds (avdp.)	220.46226
Radians	Circumferences	0.15915494
	Degrees	57.295779
	Minutes	3437.7468
	Quadrants	0.63661977
	Revolutions	0.15915494
	Seconds	206,264.81

To convert from	To	Multiply by
Radians/cm.	Degrees/cm.	57.295779
	Degrees/ft.	1746.3754
	Degrees/inch	145.53128
	Minutes/cm.	3437.7468
Radians/sec.	Degrees/sec.	57.295779
	Revolutions/min.	9.5492966
	Revolutions/sec.	0.15915494
Radians/(sec. × sec.)	Revolutions/(min. × min.)	572.95779
	Revolutions/(min. × sec.)	9.5492966
	Revolutions/(sec. × sec.)	0.15915494
Register tons	Cu. feet	100
	Cu. meters	2.8316847
Revolutions	Degrees	360
	Grades	400
	Quadrants	4
	Radians	6.2831853
Reyns	Centipoises	6.89476×10^{6}
Rhes	Poises^{-1}	1
Rods	Centimeters	502.92
	Chains (Gunter's)	0.25
	Chains (Ramden's)	0.165
	Feet	16.5
	Feet (U.S. Survey)	16.499967
	Furlongs	0.025
	Inches	198
	Links (Gunter's)	25
	Links (Ramden's)	16.5
	Meters	5.0292
	Miles (statute)	0.003125
	Perches	1
	Yards	5.5
Rods (Brit., volume)	Cu. feet	1000
	Cu. meters	28.316847
Roentgen	Coulombs/kilogram	2.58×10^{-4}
Roods (Brit.)	Acres	0.25
	Ares	10.117141
	Sq. perches	40
	Sq. yards	1210

To convert from	To	Multiply by
Ropes (Brit.)	Feet	20
	Meters	6.096
	Yards	6.6666666
Scruples (apoth.)	Drams (apoth. *or* troy)	0.333333
	Drams (avdp.)	0.73142857
	Grains	20
	Grams	1.2959782
	Ounces (apoth. *or* troy)	0.041666
	Ounces (avdp.)	0.045714286
	Pennyweights	0.833333
	Pounds (apoth. *or* troy)	0.003472222
	Pounds (avdp.)	0.0028571429
Scruples (Brit., fluid)	Minims (Brit.)	20
Seams (Brit.)	Bushels (Brit.)	8
	Cu. feet	10.27479
	Liters	290.9414
Seconds (angular)	Degrees	0.000277777
	Minutes	0.0166666
	Radians	4.8481368×10^{-6}
Seconds (mean solar)	Days (mean solar)	1.1574074×10^{-5}
	Days (sidereal)	1.1605763×10^{-5}
	Hours (mean solar)	0.0002777777
	Hours (sidereal)	0.00027853831
	Minutes (mean solar)	0.0166666
	Minutes (sidereal)	0.016712298
	Seconds (sidereal)	1.00273791
Seconds (sidereal)	Days (mean solar)	1.1542472×10^{-5}
	Days (sidereal)	1.1574074×10^{-5}
	Hours (mean solar)	0.00027701932
	Hours (sidereal)	0.000277777
	Minutes (mean solar)	0.016621159
	Minutes (sidereal)	0.0166666
	Seconds (mean solar)	0.99726957
Shakes	Seconds	1×10^{-8}
Siemen's units	(Same as mhos)	

To convert from	To	Multiply by
Skeins	Feet	360
	Meters	109.728
Slugs	Geepounds	1
	Kilograms	14.5939
	Pounds (avdp.)	32.1740
Slugs/cu. ft.	Grams/cu. cm.	0.515379
Space (entire)	Hemispheres	2
	Steradians	12.566371
Spans	Centimeters	22.86
	Fathoms	0.125
	Feet	0.75
	Inches	9
	Quarters (Brit. linear)	1
Spherical right angles	Hemispheres	0.25
	Spheres	0.125
	Steradians	1.5707963
Sq. centimeters	Ares	1×10^{-6}
	Circ. mm.	127.32395
	Circ. mils	197,352.52
	Sq. chains (Gunter's)	2.4710538×10^{-7}
	Sq. chains (Ramden's)	1.0763910×10^{-7}
	Sq. decimeters	0.01
	Sq. feet	0.0010763910
	Sq. ft. (U.S. Survey)	0.0010763867
	Sq. inches	0.15500031
	Sq. meters	0.0001
	Sq. mm.	100
	Sq. mils	155,000.31
	Sq. rods	3.9536861×10^{-6}
	Sq. yards	0.00011959900
Sq. chains (Gunter's)	Acres	0.1
	Sq. feet	4356
	Sq. ft. (U.S. Survey)	4355.9826
	Sq. inches	627,264
	Sq. links (Gunter's)	10,000
	Sq. meters	404.68564
	Sq. miles	0.00015625

To convert from	To	Multiply by
	Sq. rods	16
	Sq. yards	484
Sq. chains (Ramden's)	Acres	0.22956841
	Sq. feet	10,000
	Sq. ft. (U.S. Survey)	9999.9600
	Sq. inches	1.44×10^{6}
	Sq. links (Ramden's)	10,000
	Sq. meters	929.0304
	Sq. miles	0.00035870064
	Sq. rods	36.730946
	Sq. yards	1111.111
Sq. decimeters	Sq. cm.	100
	Sq. inches	15.500031
Square degrees	Steradians	0.00030461742
Sq. dekameters	Acres	0.024710538
	Ares	1
	Sq. meters	100
	Sq. yards	119.59900
Sq. feet	Acres	2.295684×10^{-5}
	Ares	0.0009290304
	Sq. cm.	929.0304
	Sq. chains (Gunter's)	0.00022956841
	Sq. ft. (U.S. Survey)	0.99999600
	Sq. inches	144
	Sq. links (Gunter's)	2.2956841
	Sq. meters	0.09290304
	Sq. miles	3.5870064×10^{-8}
	Sq. rods	0.0036730946
	Sq. yards	0.111111
Sq. feet (U.S. Survey)	Acres	$2.29569330 \times 10^{-5}$
	Sq. centimeters	929.03412
	Sq. chains (Ramden's)	0.00010000040
Sq. hectometers	Sq. meters	10,000
Sq. inches	Circ. mils	1,273,239.5
	Sq. cm.	6.4516
	Sq. chains (Gunter's)	1.5942251×10^{-6}
	Sq. decimeters	0.064516
	Sq. feet	0.0069444

To convert from	To	Multiply by
	Sq. ft. (U.S. Survey)	0.0069444167
	Sq. links (Gunter's)	0.01594225
	Sq. meters	0.00064516
	Sq. miles	$2.4909767 \times 10^{-10}$
	Sq. mm.	645.16
	Sq. mils	1×10^{6}
Sq. inches/sec.	Sq. cm./hr.	23,225.76
	Sq. cm./sec.	6.4516
	Sq. ft./min.	0.416666
Sq. kilometers	Acres	247.10538
	Sq. feet	1.0763910×10^{7}
	Sq. ft. (U.S. Survey)	1.0763867×10^{7}
	Sq. inches	1.5500031×10^{9}
	Sq. meters	1×10^{6}
	Sq. miles	0.38610216
	Sq. yards	1.1959900×10^{6}
Sq. links (Gunter's)	Acres	1×10^{-5}
	Sq. cm.	404.68564
	Sq. chains (Gunter's)	0.0001
	Sq. feet	0.4356
	Sq. ft. (U.S. Survey)	0.43559826
	Sq. inches	62.7264
Sq. links (Ramden's)	Acres	2.2956841×10^{-5}
	Sq. feet	1
Sq. meters	Acres	0.00024710538
	Ares	0.01
	Hectares	0.0001
	Sq. cm.	10,000
	Sq. feet	10.763910
	Sq. inches	1550.0031
	Sq. kilometers	1×10^{-6}
	Sq. links (Gunter's)	24.710538
	Sq. links (Ramden's)	10.763910
	Sq. miles	3.8610216×10^{-7}
	Sq. mm.	1×10^{6}
	Sq. rods	0.039536861
	Sq. yards	1.1959900
Sq. miles	Acres	640
	Hectares	258.99881
	Sq. chains (Gunter's)	6400

To convert from	To	Multiply by
	Sq. feet	2.7878288×10^7
	Sq. ft. (U.S. Survey)	2.78288×10^7
	Sq. kilometers	2.5899881
	Sq. meters	2,589,988.1
	Sq. rods	102,400
	Sq. yards	3.0976×10^6
Sq. millimeters	Circ. mm.	1.2732395
	Circ. mils	1973.5252
	Sq. cm.	0.01
	Sq. inches	0.0015500031
	Sq. meters	1×10^{-6}
Sq. mils	Circ. mils	1.2732395
	Sq. cm.	6.4516×10^{-6}
	Sq. inches	1×10^{-6}
	Sq. mm.	0.00064516
Sq. rods	Acres	0.00625
	Ares	0.2529285264
	Hectares	0.002529285264
	Sq. cm.	252,928.5264
	Sq. feet	272.25
	Sq. ft. (U.S. Survey)	272.24891
	Sq. inches	39,204
	Sq. links (Gunter's)	625
	Sq. links (Ramden's)	272.25
	Sq. meters	25.29285264
	Sq. miles	9.765625×10^{-6}
	Sq. yards	30.25
Sq. yards	Acres	0.00020661157
	Ares	0.0083612736
	Hectares	8.3612736×10^{-5}
	Sq. cm.	8361.2736
	Sq. chains (Gunter's)	0.0020661157
	Sq. chains (Ramden's)	0.0009
	Sq. feet	9
	Sq. ft. (U.S. Survey)	8.9999640
	Sq. inches	1296
	Sq. links (Gunter's)	20.661157
	Sq. links (Ramden's)	9
	Sq. meters	0.83612736
	Sq. miles	$3.228305785 \times 10^{-7}$

To convert from	To	Multiply by
	Sq. perches (Brit.)	0.033057851
	Sq. rods	0.033057851
Statamperes	Abamperes	3.335635×10^{-11}
	Amperes	3.335635×10^{-10}
	E.M. cgs. units of current	3.335635×10^{-11}
	E.S. cgs. units	1
Statcoulombs	Ampere-hours	9.265653×10^{-14}
	Coulombs	3.335635×10^{-10}
	Electronic charges	2.082093×10^{9}
	E. M. cgs. units of electric charge	3.335635×10^{-11}
Statfarads	E.M. cgs. units of capacitance	1.112646×10^{-21}
	E.S. cgs. units	1
	Farads	1.112646×10^{-12}
	Microfarads	1.112646×10^{-6}
Stathenries	Abhenries	8.987584×10^{20}
	E.M. cgs. units of inductance	8.987584×10^{20}
	E.S. cgs. units	1
	Henries	8.987584×10^{11}
	Millihenries	8.987584×10^{14}
Statohms	Abohms	8.987584×10^{20}
	E.S. cgs. units	1
	Ohms	8.987584×10^{11}
Statvolts	Abvolts	2.997930×10^{10}
	Volts	299.7930
Statvolts/cm.	Volts/cm.	299.7930
	Volts/inch	761.4742
Statvolts/inch	Volts/cm.	118.0287
Steradians	Hemispheres	0.15915494
	Solid angles	0.079577472
	Spheres	0.079577472
	Spher. right angles	0.63661977
	Square degrees	3282.8063
Steres	Cubic meters	1
	Decisteres	10
	Dekasteres	0.1
	Liters	999.972

To convert from	To	Multiply by
Stilbs	Candles/sq. cm.	1
	Candles/sq. inch	6.4516
	Lamberts	3.1415927
Stokes	Cgs. units of kinematic viscosity	1
	Sq. cm./sec.	1
	Sq. inches/sec.	0.15500031
	Poise cu. cm./gram	1
Stones (Brit., legal)	Centals (Brit.)	0.14
Tons (long)	Dynes	9.96402×10^8
	Hundredweights (long)	20
	Hundredweights (short)	22.4
	Kilograms	1016.0469
	Ounces (avdp.)	35,840
	Pounds (apoth. *or* troy)	2722.22
	Pounds (avdp.)	2240
	Tons (metric)	1.1060469
	Tons (short)	1.12
Tons (metric)	Dynes	9.80665×10^8
	Grams	1×10^6
	Hundredweights (short)	22.046226
	Kilograms	1000
	Ounces (avdp.)	35,273.962
	Pounds (apoth. *or* troy)	2679.2289
	Pounds (avdp.)	2204.6226
	Tons (long)	0.98420653
	Tons (short)	1.1023113
Tons (short)	Dynes	8.89644×10^8
	Hundredweights (short)	20
	Kilograms	907.18474
	Ounces (avdp.)	32,000
	Pounds (apoth. *or* troy)	2430.555
	Pounds (avdp.)	2000
	Tons (long)	0.89285714
	Tons (metric)	0.90718474
Tons of refrig. (U.S., comm.)	B.t.u. (IST.)/hr.	12,000
	B.t.u. (IST.)/min.	200
	Kilocal. (IST.)/hr.	3023.949

To convert from	To	Multiply by
	Horsepower	4.71611
	Kg. of ice melted/hr.	37.971
	Lb. of ice melted/hr.	83.711
Tons of refrig. (U.S., std.)	B.t.u. (IST.)	288,000
	B.t.u. (mean)	287,774
	Kilocal. (IST.)	72,574.8
	Kilocal. (mean)	72,517.9
	Lb. of ice melted	2009.1
Tons (long)/sq. ft.	Atmospheres	1.05849
	Dynes/sq. cm.	1.07252×10^6
	Grams/sq. cm.	1093.6638
	Pounds/sq. ft.	2240
Tons (short)/sq. ft.	Atmospheres	0.945082
	Dynes/sq. cm.	957.605
	Grams/sq. cm.	976.486
	Pounds/sq. inch	13.8888
Tons (long)/sq. in.	Atmospheres	152.423
	Dynes/sq. cm.	1.54443×10^8
	Grams/sq. cm.	157,487.59
Tons (short)/sq. in.	Dynes/sq. cm.	1.37895×10^8
	Kg./sq. mm.	1406.139
	Pounds/sq. inch	2000
Torrs (*or* Tors)	Millimeters of Hg (0°C.)	1
	Pascals	133.3224
Townships (U.S.)	Acres	23,040
	Sections	36
	Sq. miles	36
Tuns	Gallons (U.S.)	252
	Hogsheads	4
Volts	Abvolts	1×10^8
	Mks. (r *or* nr) units	1
	Statvolts	0.003335635
	Volts (Int.)	0.999670
Volts (Int.)	Volts	1.000330
Volts/°C.	Joules/(coulomb × °C.)	1
Volt-coulombs	Joules(Int.)	0.999835
Volt-coulombs (Int.)	Joules	1.000165

To convert from	To	Multiply by
Volt-electronic charge-seconds	Planck's constant	2.41814×10^{14}
Volt-faraday (chem.)-seconds	Planck's constant	1.45650×10^{38}
Volt-faraday (phys.)-seconds	Planck's constant	1.45690×10^{38}
Volt-seconds	Maxwells	1×10^{8}
Watts	B.t.u./hr.	3.41443
	B.t.u. (mean)/hr.	3.40952
	B.t.u. (mean)/min.	0.0568253
	B.t.u./sec.	0.000948451
	B.t.u. (mean)/sec.	0.000947088
	Cal./hr.	860.421
	Cal. (mean)/hr.	859.184
	Cal. (20°C.)/hr.	860.853
	Cal./min.	14.3403
	Cal. (IST.)/min.	14.3310
	Cal. (mean)/min.	14.3197
	Cal., *kh.*/min.	0.0143403
	Kilocal. (IST.)/min.	0.0143310
	Kilocal. (mean)/min.	0.0143197
	Ergs/sec.	1×10^{7}
	Foot-pounds/min.	44.2537
	Horsepower	0.00134102
	Horsepower (boiler)	0.000101942
	Horsepower (elec.)	0.00134048
	Horsepower (metric)	0.00135962
	Joules/sec.	1
	Kilowatts	0.001
	Liter-atm./hr.	35.5282
Watts (Int.)	B.t.u./hr.	3.41499
	B.t.u. (mean)/hr.	3.41008
	B.t.u./min.	0.569165
	B.t.u. (mean)/min.	0.0568347
	Cal./hr.	860.563
	Cal. (mean)/hr.	859.326
	Kilocal./min.	0.0143427
	Kilocal. (IST.)/min.	0.0143333

To convert from	To	Multiply by
	Kilocal. (mean)/min.	0.0143221
	Ergs/sec.	1.000165×10^7
	Joules (Int.)/sec.	1
	Watts	1.000165
Watts/sq. cm.	B.t.u./(hr. × sq. ft.)	3172.10
	Cal./(hr. × sq. cm.)	860.421
	Ft.-lb./(min. × sq. ft.)	41,113.1
Watts/sq. in.	B.t.u./(hr. × sq. ft.)	491.677
	Cal./(hr. × sq. cm.)	133.365
	Ft.-lb./(min. × sq. ft.)	6372.54
Watt-hours	B.t.u.	3.41443
	B.t.u. (mean)	3.40952
	Cal.	860.421
Watt-hours	Kilocal. (mean)	0.859184
	Cal. (mean)	859.184
	Foot-pounds	2655.22
	Hp.-hours	0.00134102
	Joules	3600
	Joules (Int.)	3599.41
	Kg.-meters	367.098
	Kw.-hours	0.001
	Watt-hours (Int.)	0.999835
Watt-sec.	Foot-pounds	0.737562
	Gram-cm.	10,197.16
	Joules	1
	Liter-atm.	0.00986895
	Volt-coulombs	1
Wavelength of orange-red line of krypton 86	Ångström units	6057.80211
	Millimeters	0.000605780211
Wavelength of red line of cadmium	Ångström units	6438.4696
	Millimeters	0.00064384696
Webers	Cgs. units of induction flux	1×10^8
	E.M. cgs. units of induction flux	1×10^8
	Lines	1×10^8
	Maxwells	1×10^8
	Mks. units of induction flux	1

To convert from	To	Multiply by
	Mks. units of magnetic charge	0.079577472
	Mks. units of magnetic charge	1
	Volt-seconds	1
Webers/sq. cm.	Gausses	1×10^8
	Lines	1×10^8
	Lines/sq. inch	6.4516×10^8
Webers/sq. in.	Gausses	1.5500031×10^7
Weeks (mean calendar)	Days (mean solar)	7
	Days (sidereal)	7.0191654
	Hours (mean solar)	168
	Hours (sidereal)	168.45997
	Minutes (mean solar)	10,080
	Minutes (sidereal)	10,107.598
	Months (lunar)	0.23704235
	Months (mean calendar)	0.23013699
	Years (calendar)	0.019178082
	Years (sidereal)	0.019164622
	Years (tropical)	0.019165365
Weys (Brit., mass.)	Pounds (avdp.)	252
X units	Meters	1.00202×10^{-13}
Yards	Centimeters	91.44
	Chains (Gunter's)	0.4545454
	Chains (Ramden's)	0.03
	Cubits	2
	Fathoms	0.5
	Feet	3
	Feet (U.S. Survey)	2.9999940
	Furlongs	0.00454545
	Inches	36
	Meters	0.9144
	Poles (Brit.)	0.181818
	Quarters (Brit., linear)	4
	Rods	0.181818
	Spans	4

To convert from	To	Multiply by
Years (calendar)	Days (mean solar)	365
	Hours (mean solar)	8760
	Minutes (mean solar)	525,600
	Months (lunar)	12.360065
	Months (mean calendar)	12
	Seconds (mean solar)	3.1536×10^7
	Weeks (mean calendar)	52.142857
	Years (sidereal)	0.99929814
	Years (tropical)	0.99933690
Years (leap)	Days (mean solar)	366
Years (sidereal)	Days (mean solar)	365.25636
	Days (sidereal)	366.25640
	Years (calendar)	1.0007024
	Years (tropical)	1.0000388
Years (tropical)	Days (mean solar)	365.24219
	Days (sidereal)	366.24219
	Hours (mean solar)	8765.8126
	Hours (sidereal)	8789.8126
	Months (mean calendar)	12.007963
	Seconds (mean solar)	3.1556926×10^7
	Seconds (sidereal)	3.1643326×10^7
	Weeks (mean calendar)	52.177456
	Years (calendar)	1.0006635
	Years (sidereal)	0.99996121

Temperature Conversion Tables

The following tables are derived from the *Smithsonian Metrological Tables*, Sixth Revised Edition, Fifth Reprint, issued 1971.

Approximate Absolute, Centigrade, Fahrenheit, and Reaumur Temperature Scales

Scale	Symbol	Freezing point of water (1 atmos.)	Boiling point of water (1 atmos.)	Conversion formulae
Centigrade*	°C.	0	100	C = (5/9) (F − 32) = (5/4) R = K − 273.16 = AA − 273
Fahrenheit	°F.	32	212	F = (9/5) C + 32 = (9/4) R + 32 = (9/5) (K − 273.16) + 32
Reaumur	°R.	0	80	R = (4/9) (F − 32) = (4/5) C = (4/5) (K − 273.16)
Thermodynamic Kelvin / Absolute Centigrade	K, A	273.16 ± 0.01‡	373.16 ± 0.01†	K = C + 273.16 = AA + 0.16 = (5/9) (F − 32) + 273.16
Approximate Absolute	AA.	273	373	AA = C + 273 = K − 0.16 = (5/9) (F − 32) + 273
Rankine / Absolute Fahrenheit	—	491.69	671.69	Rankine = F + 459.69

*The Ninth General Conference on Weights and Measures, October 1948, gave the degree of temperature the designation of *degree Celsius* in place of *degree centigrade*. See Stimson, H. F., The international temperature scale of 1948, Nat. Bur. Stand. Journ. Res., vol. 42, p. 209, 1949, and Amer. Journ. Phys., vol. 23, p. 614, 1955.

†R. T. Birge, Rev. Mod. Phys., vol. 13, p. 233, 1941.

‡In 1954, the thermodynamic temperature was defined so that 273.16°K corresponds to the triple point, yielding the value 273.15°K as equivalent to 0°C. Dixieme Conférence Générale Poids et Mesures, Compt. Rend., 1954.

Proportional Parts

C, K, AA	0.1	0.2	0.3	0.4	0.5	0.6	0.7	0.8	0.9
F	0.18	0.36	0.54	0.72	0.90	1.08	1.26	1.44	1.62
R	0.08	0.16	0.24	0.32	0.40	0.48	0.56	0.64	0.72
F	0.055+	0.111+	0.166+	0.222+	0.277+	0.333+	0.388+	0.444+	0.5
C, K, AA	0.044+	0.088+	0.133+	0.177+	0.222+	0.266+	0.311+	0.355+	0.4
R	0.1	0.2	0.3	0.4	0.5	0.6	0.7	0.8	0.9
R	0.1	0.2	0.3	0.4	0.5	0.6	0.7	0.8	0.9
C, K, AA	0.125	0.25	0.375	0.5	0.625	0.75	0.875	1.0	1.125
F	0.225	0.45	0.675	0.9	1.125	1.35	1.575	1.8	2.025

AA	C	F	R	AA	C	F	R	AA	C	F	R
375	102	215.6	81.6	350	77	170.6	61.6	325	52	125.6	41.6
374	101	213.8	80.8	349	76	168.8	60.8	324	51	123.8	40.8
373	100	212.0	80.0	348	75	167.0	60.0	323	50	122.0	40.0
372	99	210.2	79.2	347	74	165.2	59.2	322	49	120.2	39.2
371	98	208.4	78.4	346	73	163.4	58.4	321	48	118.4	38.4
370	97	206.6	77.6	345	72	161.6	57.6	320	47	116.6	37.6
369	96	204.8	76.8	344	71	159.8	56.8	319	46	114.8	36.8
368	95	203.0	76.0	343	70	158.0	56.0	318	45	113.0	36.0
367	94	201.2	75.2	342	69	156.2	55.2	317	44	111.2	35.2
366	93	199.4	74.4	341	68	154.4	54.4	316	43	109.4	34.4
365	92	197.6	73.6	340	67	152.6	53.6	315	42	107.6	33.6
364	91	195.8	72.8	339	66	150.8	52.8	314	41	105.8	32.8
363	90	194.0	72.0	338	65	149.0	52.0	313	40	104.0	32.0
362	89	192.2	71.2	337	64	147.2	51.2	312	39	102.2	31.2
361	88	190.4	70.4	336	63	145.4	50.4	311	38	100.4	30.4

AA	C	F	R	AA	C	F	R	AA	C	F	R
360	87	188.6	69.6	335	62	143.6	49.6	310	37	98.6	29.6
359	86	186.8	68.8	334	61	141.8	48.8	309	36	96.8	28.8
358	85	185.0	68.0	333	60	140.0	48.0	308	35	95.0	28.0
357	84	183.2	67.2	332	59	138.2	47.2	307	34	93.2	27.2
356	83	181.4	66.4	331	58	136.4	46.4	306	33	91.4	26.4
355	82	179.6	65.6	330	57	134.6	45.6	305	32	89.6	25.6
354	81	177.8	64.8	329	56	132.8	44.8	304	31	87.8	24.8
353	80	176.0	64.0	328	55	131.0	44.0	303	30	86.0	24.0
352	79	174.2	63.2	327	54	129.2	43.2	302	29	84.2	23.2
351	78	172.4	62.4	326	53	127.4	42.4	301	28	82.4	22.4
350	77	170.6	61.6	325	52	125.6	41.6	300	27	80.6	21.6
300	27	80.6	21.6	250	−23	−9.4	−18.4	200	−73	−99.4	−58.4
299	26	78.8	20.8	249	24	11.2	19.2	199	74	101.2	59.2
298	25	77.0	20.0	248	25	13.0	20.0	198	75	103.0	60.0
297	24	75.2	19.2	247	26	14.8	20.8	197	76	104.8	60.8
296	23	73.4	18.4	246	27	16.6	21.6	196	77	106.6	61.6
295	22	71.6	17.6	245	−28	−18.4	−22.4	195	−78	−108.4	−62.4
294	21	69.8	16.8	244	29	20.2	23.2	194	79	110.2	63.2
293	20	68.0	16.0	243	30	22.0	24.0	193	80	112.0	64.0
292	19	66.2	15.2	242	31	23.8	24.8	192	81	113.8	64.8
291	18	64.4	14.4	241	32	25.6	25.6	191	82	115.6	65.6
290	17	62.6	13.6	240	−33	−27.4	−26.4	190	−83	−117.4	−66.4
289	16	60.8	12.8	239	34	29.2	27.2	189	84	119.2	67.2
288	15	59.0	12.0	238	35	31.0	28.0	188	85	121.0	68.0
287	14	57.2	11.2	237	36	32.8	28.8	187	86	122.8	68.8
286	13	55.4	10.4	236	37	34.6	29.6	186	87	124.6	69.6

285	12	53.6	9.6	235	−38	−36.4	−30.4	185	−88	−126.4	−70.4
284	11	51.8	8.8	234	39	38.2	31.2	184	89	128.2	71.2
283	10	50.0	8.0	233	40	40.0	32.0	183	90	130.0	72.0
282	9	48.2	7.2	232	41	41.8	32.8	182	91	131.8	72.8
281	8	46.4	6.4	231	42	43.6	33.6	181	92	133.6	73.6
280	7	44.6	5.6	230	−43	−45.4	−34.4	180	−93	−135.4	−74.4
279	6	42.8	4.8	229	44	47.2	35.2	179	94	137.2	75.2
278	5	41.0	4.0	228	45	49.0	36.0	178	95	139.0	76.0
277	4	39.2	3.2	227	46	50.8	36.8	177	96	140.8	76.8
276	3	37.4	2.4	226	47	52.6	37.6	176	97	142.6	77.6
275	+2	35.6	+1.6	225	−48	−54.4	−38.4	175	−98	−144.4	−78.4
274	+1	33.8	+0.8	224	49	56.2	39.2	174	99	146.2	79.2
273	0	32.0	0.0	223	50	58.0	40.0	173	100	148.0	80.0
272	−1	30.2	−0.8	222	51	59.8	40.8	172	101	149.8	80.8
271	−2	28.4	−1.6	221	52	61.6	41.6	171	102	151.6	81.6
270	−3	26.6	−2.4	220	−53	−63.4	−42.4	170	−103	−153.4	−82.4
269	4	24.8	3.2	219	54	65.2	43.2	169	104	155.2	83.2
268	5	23.0	4.0	218	55	67.0	44.0	168	105	157.0	84.0
267	6	21.2	4.8	217	56	68.8	44.8	167	106	158.8	84.8
266	7	19.4	5.6	216	57	70.6	45.6	166	107	160.6	85.6
265	−8	17.6	−6.4	215	−58	−72.4	−46.4	165	−108	−162.4	−86.4
264	9	15.8	7.2	214	59	74.2	47.2	164	109	164.2	87.2
263	10	14.0	8.0	213	60	76.0	48.0	163	110	166.0	88.0
262	11	12.2	8.8	212	61	77.8	48.8	162	111	167.8	88.8
261	12	10.4	9.6	211	62	79.6	49.6	161	112	169.6	89.6
260	−13	8.6	−10.4	210	−63	−81.4	−50.4	160	−113	−171.4	−90.4
259	14	6.8	11.2	209	64	83.2	51.2	159	114	173.2	91.2
258	15	5.0	12.0	208	65	85.0	52.0	158	115	175.0	92.0
257	16	3.2	12.8	207	66	86.8	52.8	157	116	176.8	92.8
256	17	+1.4	13.6	206	67	88.6	53.6	156	117	178.6	93.6

AA	C	F	R	AA	C	F	R	AA	C	F	R
255	−18	−0.4	−14.4	205	−68	−90.4	−54.4	155	−118	−180.4	−94.4
254	19	2.2	15.2	204	69	92.2	55.2	154	119	182.2	95.2
253	20	4.0	16.0	203	70	94.0	56.0	153	120	184.0	96.0
252	21	5.8	16.8	202	71	95.8	56.8	152	121	185.8	96.8
251	22	7.6	17.6	201	72	97.6	57.6	151	122	187.6	97.6
250	−23	−9.4	−18.4	200	−73	−99.4	−58.4	150	−123	−189.4	−98.4
150	−123	−189.4	−98.4	100	−173	−279.4	−138.4	50	−223	−369.4	−178.4
149	124	191.2	99.2	99	174	281.2	139.2	49	224	371.2	179.2
148	125	193.0	100.0	98	175	283.0	140.0	48	225	373.0	180.0
147	126	194.8	100.8	97	176	284.8	140.8	47	226	374.8	180.8
146	127	196.6	101.6	96	177	286.6	141.6	46	227	376.6	181.6
145	−128	−198.4	−102.4	95	−178	−288.4	−142.4	45	−228	−378.4	−182.4
144	129	200.2	103.2	94	179	290.2	143.2	44	229	380.2	183.2
143	130	202.0	104.0	93	180	292.0	144.0	43	230	382.0	184.0
142	131	203.8	104.8	92	181	293.8	144.8	42	231	383.8	184.8
141	132	205.6	105.6	91	182	295.6	145.6	41	232	385.6	185.6
140	−133	−207.4	−106.4	90	−183	−297.4	−146.4	40	−233	−387.4	−186.4
139	134	209.2	107.2	89	184	299.2	147.2	39	234	389.2	187.2
138	135	211.0	108.0	88	185	301.0	148.0	38	235	391.0	188.0
137	136	212.8	108.8	87	186	302.8	148.8	37	236	392.8	188.8
136	137	214.6	109.6	86	187	304.6	149.6	36	237	394.6	189.6
135	−138	−216.4	−110.4	85	−188	−306.4	−150.4	35	−238	−396.4	−190.4
134	139	218.2	111.2	84	189	308.2	151.2	34	239	398.2	191.2
133	140	220.0	112.0	83	190	310.0	152.0	33	240	400.0	192.0
132	141	221.8	112.8	82	191	311.8	152.8	32	241	401.8	192.8
131	142	223.6	113.6	81	192	313.6	153.6	31	242	403.6	193.6
130	−143	−225.4	−114.4	80	−193	−315.4	−154.4	30	−243	−405.4	−194.4
129	144	227.2	115.2	79	194	317.2	155.2	29	244	407.2	195.2
128	145	229.0	116.0	78	195	319.0	156.0	28	245	409.0	196.0

127	146	230.8	116.8	77	196	320.8	156.8	27	246	410.8	196.8
126	147	232.6	117.6	76	197	322.6	157.6	26	247	412.6	197.6
125	−148	−234.4	−118.4	75	−198	−324.4	−158.4	25	−248	−414.4	−198.4
124	149	236.2	119.2	74	199	326.2	159.2	24	249	416.2	199.2
123	150	238.0	120.0	73	200	328.0	160.0	23	250	418.0	200.0
122	151	239.8	120.8	72	201	329.8	160.8	22	251	419.8	200.8
121	152	241.6	121.6	71	202	331.6	161.6	21	252	421.6	201.6
120	−153	−243.4	−122.4	70	−203	−333.4	−162.4	20	−253	−423.4	−202.4
119	154	245.2	123.2	69	204	335.2	163.2	19	254	425.2	203.2
118	155	247.0	124.0	68	205	337.0	164.0	18	255	427.0	204.0
117	156	248.8	124.8	67	206	338.8	164.8	17	256	428.8	204.8
116	157	250.6	125.6	66	207	340.6	165.6	16	257	430.6	205.6
115	−158	−252.4	−126.4	65	−208	−342.4	−166.4	15	−258	−432.4	−206.4
114	159	254.2	127.2	64	209	344.2	167.2	14	259	434.2	207.2
113	160	256.0	128.0	63	210	346.0	168.0	13	260	436.0	208.0
112	161	257.8	128.8	62	211	347.8	168.8	12	261	437.8	208.8
111	162	259.6	129.6	61	212	349.6	169.6	11	262	439.6	209.6
110	−163	−261.4	−130.4	60	−213	−351.4	−170.4	10	−263	−441.4	−210.4
109	164	263.2	131.2	59	214	353.2	171.2	9	264	443.2	211.2
108	165	265.0	132.0	58	215	355.0	172.0	8	265	445.0	212.0
107	166	266.8	132.8	57	216	356.8	172.8	7	266	446.8	212.8
106	167	268.6	133.6	56	217	358.6	173.6	6	267	448.6	213.6
105	−168	−270.4	−134.4	55	−218	−360.4	−174.4	5	−268	−450.4	−214.4
104	169	272.2	135.2	54	219	362.2	175.2	4	269	452.2	215.2
103	170	274.0	136.0	53	220	364.0	176.0	3	270	454.0	216.0
102	171	275.8	136.8	52	221	365.8	176.8	2	271	455.8	216.8
101	172	277.6	137.6	51	222	367.6	177.6	1	272	457.6	217.6
100	−173	−279.4	−138.4	50	−223	−369.4	−178.4	0	−273	−459.4	−218.4

Fahrenheit to Centigrade

Fahrenheit	.0°C.	.1°C.	.2°C.	.3°C.	.4°C.	.5°C.	.6°C.	.7°C.	.8°C.	.9°C.
+130	+54.44	+54.50	+54.56	+54.61	+54.67	+54.72	+54.78	+54.83	+54.89	+54.94
129	53.89	53.94	54.00	54.06	54.11	54.17	54.22	54.28	54.33	54.39
128	53.33	53.29	53.44	53.50	53.56	53.61	53.67	53.72	53.78	53.83
127	52.78	52.83	52.89	52.94	53.00	53.06	53.11	53.17	53.22	53.28
126	52.22	52.28	52.33	52.39	52.44	52.50	52.56	52.61	52.67	52.72
+125	+51.67	+51.72	+51.78	+51.83	+51.89	+51.94	+52.00	+52.06	+52.11	+52.17
124	51.11	51.17	51.22	51.28	51.33	51.39	51.44	51.50	51.56	51.61
123	50.56	50.61	50.67	50.72	50.78	50.83	50.89	50.94	51.00	51.06
122	50.00	50.06	50.11	50.17	50.22	50.28	50.33	50.39	50.44	50.50
121	49.44	49.50	49.56	49.61	49.67	49.72	49.78	49.83	49.89	49.94
+120	+48.89	+48.94	+49.00	+49.06	+49.11	+49.17	+49.22	+49.28	+49.33	+49.39
119	48.33	48.39	48.44	48.50	48.56	48.61	48.67	48.72	48.78	48.83
118	47.78	47.83	47.89	47.94	48.00	48.06	48.11	48.17	48.22	48.28
117	47.22	47.28	47.33	47.39	47.44	47.50	47.56	47.61	47.67	47.72
116	46.67	46.72	46.78	46.83	46.89	46.94	47.00	47.06	47.11	47.17
+115	+46.11	+46.17	+46.22	+46.28	+46.33	+46.39	+46.44	+46.50	+46.56	+46.61
114	45.56	45.61	45.67	45.72	45.78	45.83	45.89	45.94	46.00	46.06
113	45.00	45.06	45.11	45.17	45.22	45.28	45.33	45.39	45.44	45.50
112	44.44	44.50	44.56	44.61	44.67	44.72	44.78	44.83	44.89	44.94
111	43.89	43.94	44.00	44.06	44.11	44.17	44.22	44.28	44.33	44.39
+110	+43.33	+43.39	+43.44	+43.50	+43.56	+43.61	+43.67	+43.72	+43.78	+43.83
109	42.78	42.83	42.89	42.94	43.00	43.06	43.11	43.17	43.22	43.28
108	42.22	42.28	42.33	42.39	42.44	42.50	42.56	42.61	42.67	42.72
107	41.67	41.72	41.78	41.83	41.89	41.94	42.00	42.06	42.11	42.17
196	41.11	41.17	41.22	41.28	41.33	41.39	41.44	41.50	41.56	41.61
+105	+40.56	+40.61	+40.67	+40.72	+40.78	+40.83	+40.89	+40.94	+41.00	+41.06
104	40.00	40.06	40.11	40.17	40.22	40.28	40.33	40.39	40.44	40.50
103	39.44	39.50	39.56	39.61	39.67	39.72	39.78	39.83	39.89	39.94
102	38.89	38.94	39.00	39.06	39.11	39.17	39.22	39.28	39.33	39.39
101	38.33	38.39	38.44	38.50	38.56	38.61	38.67	38.72	38.78	38.83
+100	+37.78	+37.83	+37.89	+37.94	+38.00	+38.06	+38.11	+38.17	+38.22	+38.28
99	37.22	37.28	37.33	37.39	37.44	37.50	37.56	37.61	37.67	37.7
98	36.67	36.72	36.78	36.83	36.89	36.94	37.00	37.06	37.11	37.1
97	36.11	36.17	36.22	36.28	36.33	36.39	36.44	36.50	36.56	36.6
96	35.56	35.61	35.67	35.72	35.78	35.83	35.89	35.94	36.00	36.0
+95	+35.00	+35.06	+35.11	+35.17	+35.22	+35.28	+35.33	+35.39	+35.44	+35.5
94	34.44	34.50	34.56	34.61	34.67	34.72	34.78	34.83	34.89	34.9
93	33.89	33.94	34.00	34.06	34.11	34.17	34.22	34.28	34.33	34.3
92	33.33	33.39	33.44	33.50	33.56	33.61	33.67	33.72	33.78	33.8
91	32.78	32.83	32.89	32.94	33.00	33.06	33.11	33.17	33.22	33.2

Fahrenheit	.0°C.	.1°C.	.2°C.	.3°C	.4°C.	.5°C.	.6°C.	.7°C.	.8°C.	.9°C.
+90	+32.22	+32.28	+32.33	+32.39	+32.44	+32.50	+32.56	+32.61	+32.67	+32.72
89	31.67	31.72	31.78	31.83	31.89	31.94	32.00	32.06	32.11	32.17
88	31.11	31.17	31.22	31.28	31.33	31.39	31.44	31.50	31.56	31.61
87	30.56	30.61	30.67	30.72	30.78	30.83	30.89	30.94	31.00	31.06
86	30.00	30.06	30.11	30.17	30.22	30.28	30.33	30.39	30.44	30.50
+85	+29.44	+29.50	+29.56	+29.61	+29.67	+29.72	+29.78	+29.83	+29.89	+29.94
84	28.89	28.94	29.00	29.06	29.11	29.17	29.22	29.28	29.33	29.39
83	28.33	28.39	28.44	28.50	28.56	28.61	28.67	28.72	28.78	28.83
82	27.78	27.83	27.89	27.94	28.00	28.06	28.11	28.17	28.22	28.28
81	27.22	27.28	27.33	27.39	27.44	27.50	27.56	27.61	27.67	27.72
+80	+26.67	+26.72	+26.78	+26.83	+26.89	+26.94	+27.00	+27.06	+27.11	+27.17
79	26.11	26.17	26.22	26.28	26.33	26.39	26.44	26.50	26.56	26.61
78	25.56	25.61	25.67	25.72	25.78	25.83	25.89	25.94	26.00	26.06
77	25.00	25.06	25.11	25.17	25.22	25.28	25.33	25.39	25.44	25.50
76	24.44	24.50	24.56	24.61	24.67	24.72	24.78	24.83	24.89	24.94
+75	+23.89	+23.94	+24.00	+24.06	+24.11	+24.17	+24.22	+24.28	+24.33	+24.39
74	23.33	23.39	23.44	23.50	23.56	23.61	23.67	23.72	23.78	23.83
73	22.78	22.83	22.89	22.94	23.00	23.06	23.11	23.17	23.22	23.28
72	22.22	22.28	22.33	22.39	22.44	22.50	22.56	22.61	22.67	22.72
71	21.67	21.72	21.78	21.83	21.89	21.94	22.00	22.06	22.11	22.17
+70	+21.11	+21.17	+21.22	+21.28	+21.33	+21.39	+21.44	+21.50	+21.56	+21.61
69	20.56	20.61	20.67	20.72	20.78	20.83	20.89	20.94	21.00	21.06
68	20.00	20.06	20.11	20.17	20.22	20.28	20.33	20.39	20.44	20.50
67	19.44	19.50	19.56	19.61	19.67	19.72	19.78	19.83	19.89	19.94
66	18.89	18.94	19.00	19.06	19.11	19.17	19.22	19.28	19.33	19.39
+65	+18.33	+18.39	+18.44	+18.50	+18.56	+18.61	+18.67	+18.72	+18.78	+18.83
64	17.78	17.83	17.89	17.94	18.00	18.06	18.11	18.17	18.22	18.28
63	17.22	17.28	17.33	17.39	17.44	17.50	17.56	17.61	17.67	17.72
62	16.67	16.72	16.78	16.83	16.89	16.94	17.00	17.06	17.11	17.17
61	16.11	16.17	16.22	16.28	16.33	16.39	16.44	16.50	16.56	16.61
+60	+15.56	+15.61	+15.67	+15.72	+15.78	+15.83	+15.89	+15.94	+16.00	+16.06
59	15.00	15.06	15.11	15.17	15.22	15.28	15.33	15.39	15.44	15.50
58	14.44	14.50	14.56	14.61	14.67	14.72	14.78	14.83	14.89	14.94
57	13.89	13.94	14.00	14.06	14.11	14.17	14.22	14.28	14.33	14.39
56	13.33	13.39	13.44	13.50	13.56	13.61	13.67	13.72	13.78	13.83
+55	+12.78	+12.83	+12.89	+12.94	+13.00	+13.06	+13.11	+13.17	+13.22	+13.28
54	12.22	12.28	12.33	12.39	12.44	12.50	12.56	12.61	12.67	12.72
53	11.67	11.72	11.78	11.83	11.89	11.94	12.00	12.06	12.11	12.17
52	11.11	11.17	11.22	11.28	11.33	11.39	11.44	11.50	11.56	11.61
51	10.56	10.61	10.67	10.72	10.78	10.83	10.89	10.94	11.00	11.06

Fahrenheit	.0°C.	.1°C.	.2°C.	.3°C.	.4°C.	.5°C.	.6°C.	.7°C.	.8°C.	.9°C.
+50	+10.00	+10.06	+10.11	+10.17	+10.22	+10.28	+10.33	+10.39	+10.44	+10.50
49	9.44	9.50	9.56	9.61	9.67	9.72	9.78	9.83	9.89	9.94
48	8.89	8.94	9.00	9.06	9.11	9.17	9.22	9.28	9.33	9.39
47	8.33	8.39	8.44	8.50	8.56	8.61	8.67	8.72	8.78	8.83
46	7.78	7.83	7.89	7.94	8.00	8.06	8.11	8.17	8.22	8.28
+45	+7.22	+7.28	+7.33	+7.39	+7.44	+7.50	+7.56	+7.61	+7.67	+7.72
44	6.67	6.72	6.78	6.83	6.89	6.94	7.00	7.06	7.11	7.17
43	6.11	6.17	6.22	6.28	6.33	6.39	6.44	6.50	6.56	6.61
42	5.56	5.61	5.67	5.72	5.78	5.83	5.89	5.94	6.00	6.06
41	5.00	5.06	5.11	5.17	5.22	5.28	5.33	5.39	5.44	5.50
+40	+4.44	+4.50	+4.56	+4.61	+4.67	+4.72	+4.78	+4.83	+4.89	+4.94
39	3.89	3.94	4.00	4.06	4.11	4.17	4.22	4.28	4.33	4.39
38	3.33	3.39	3.44	3.50	3.56	3.61	3.67	3.72	3.78	3.83
37	2.78	2.83	2.89	2.94	3.00	3.06	3.11	3.17	3.22	3.28
36	2.22	2.28	2.33	2.39	2.44	2.50	2.56	2.61	2.67	2.72
+35	+1.67	+1.72	+1.78	+1.83	+1.89	+1.94	+2.00	+2.06	+2.11	+2.17
34	+1.11	+1.17	+1.22	+1.28	+1.33	+1.39	+1.44	+1.50	+1.56	+1.61
33	+0.56	+0.61	+0.67	+0.72	+0.78	+0.83	+0.89	+0.94	+1.00	+1.06
32	0.00	+0.06	+0.11	+0.17	+0.22	+0.28	+0.33	+0.39	+0.44	+0.50
31	−0.56	−0.50	−0.44	−0.39	−0.33	−0.28	−0.22	−0.17	−0.11	−0.06
+30	−1.11	−1.06	−1.00	−0.94	−0.89	−0.83	−0.78	−0.72	−0.67	−0.61
29	1.67	1.61	1.56	1.50	1.44	1.39	1.33	1.28	1.22	1.17
28	2.22	2.17	2.11	2.06	2.00	1.94	1.89	1.83	1.78	1.72
27	2.78	2.72	2.67	2.61	2.56	2.50	2.44	2.39	2.33	2.28
26	3.33	3.28	3.22	3.17	3.11	3.06	3.00	2.94	2.89	2.83
+25	−3.89	−3.83	−3.78	−3.72	−3.67	−3.61	−3.56	−3.50	−3.44	−3.39
24	4.44	4.39	4.33	4.28	4.22	4.17	4.11	4.06	4.00	3.94
23	5.00	4.94	4.89	4.83	4.78	4.72	4.67	4.61	4.56	4.50
22	5.56	5.50	5.44	5.39	5.33	5.28	5.22	5.17	5.11	5.06
21	6.11	6.06	6.00	5.94	5.89	5.83	5.78	5.72	5.67	5.61
+20	−6.67	−6.61	−6.56	−6.50	−6.44	−6.39	−6.33	−6.28	−6.22	−6.17
19	7.22	7.17	7.11	7.06	7.00	6.94	6.89	6.83	6.78	6.72
18	7.78	7.72	7.67	7.61	7.56	7.50	7.44	7.39	7.33	7.28
17	8.33	8.28	8.22	8.17	8.11	8.06	8.00	7.94	7.89	7.83
16	8.89	8.83	8.78	8.72	8.67	8.61	8.56	8.50	8.44	8.39
+15	−9.44	−9.39	−9.33	−9.28	−9.22	−9.17	−9.11	−9.06	−9.00	−8.9
14	10.00	9.94	9.89	9.83	9.78	9.72	9.67	9.61	9.56	9.5
13	10.56	10.50	10.44	10.39	10.33	10.28	10.22	10.17	10.11	10.0
12	11.11	11.06	11.00	10.94	10.89	10.83	10.78	10.72	10.67	10.6
11	11.67	11.61	11.56	11.50	11.44	11.39	11.33	11.28	11.22	11.1

Fahrenheit	.0°C.	.1°C.	.2°C.	.3°C.	.4°C.	.5°C.	.6°C.	.7°C.	.8°C.	.9°C.
+10	−12.22	−12.17	−12.11	−12.06	−12.00	−11.94	−11.89	−11.83	−11.78	−11.72
9	12.78	12.72	12.67	12.61	12.56	12.50	12.44	12.39	12.33	12.28
8	13.33	13.28	13.22	13.17	13.11	13.06	13.00	12.94	12.89	12.83
7	13.89	13.83	13.78	13.72	13.67	13.61	13.56	13.50	13.44	13.39
6	14.44	14.39	14.33	14.28	14.22	14.17	14.11	14.06	14.00	13.94
+5	−15.00	−14.94	−14.89	−14.83	−14.78	−14.72	−14.67	−14.61	−14.56	−14.50
4	15.56	15.50	15.44	15.39	15.33	15.28	15.22	15.17	15.11	15.06
3	16.11	16.06	16.00	15.94	15.89	15.83	15.78	15.72	15.67	15.61
2	16.67	16.61	16.56	16.50	16.44	16.39	16.33	16.28	16.22	16.17
1	17.22	17.17	17.11	17.06	17.00	16.94	16.89	16.83	16.78	16.72
+0	17.78	17.72	17.67	17.61	17.56	17.50	17.44	17.39	17.33	17.28
−0	−17.78	−17.83	−17.89	−17.94	−18.00	−18.06	−18.11	−18.17	−18.22	−18.28
1	18.33	18.39	18.44	18.50	18.56	18.61	18.67	18.72	18.78	18.83
2	18.89	18.94	19.00	19.06	19.11	19.17	19.22	19.28	19.33	19.39
3	19.44	19.50	19.56	19.61	19.67	19.72	19.78	19.83	19.89	19.94
4	20.00	20.06	20.11	20.17	20.22	20.28	20.33	20.39	20.44	20.50
−5	−20.56	−20.61	−20.67	−20.72	−20.78	−20.83	−20.89	−20.94	−21.00	−21.06
6	21.11	21.17	21.22	21.28	21.33	21.39	21.44	21.50	21.56	21.61
7	21.67	21.72	21.78	21.83	21.89	21.94	22.00	22.06	22.11	22.17
8	22.22	22.28	22.33	22.39	22.44	22.50	22.56	22.61	22.67	22.72
9	22.78	22.83	22.89	22.94	23.00	23.06	23.11	23.17	23.22	23.28
−10	−23.33	−23.39	−23.44	−23.50	−23.56	−23.61	−23.67	−23.72	−23.78	−23.83
11	23.89	23.94	24.00	24.06	24.11	24.17	24.22	24.28	24.33	24.39
12	24.44	24.50	24.56	24.61	24.67	24.72	24.78	24.83	24.89	24.94
13	25.00	25.06	25.11	25.17	25.22	25.28	25.33	25.39	25.44	25.50
14	25.56	25.61	25.67	25.72	25.78	25.83	25.89	25.94	26.00	26.06
−15	−26.11	−26.17	−26.22	−26.28	−26.33	−26.39	−26.44	−26.50	−26.56	−26.61
16	26.67	26.72	26.78	26.83	26.89	26.94	27.00	27.06	27.11	27.17
17	27.22	27.28	27.33	27.39	27.44	27.50	27.56	27.61	27.67	27.72
18	27.78	27.83	27.89	27.94	28.00	28.06	28.11	28.17	28.22	28.28
19	28.33	28.39	28.44	28.50	28.56	28.61	28.67	28.72	28.78	28.83
−20	−28.89	−28.94	−29.00	−29.06	−29.11	−29.17	−29.22	−29.28	−29.33	−29.39
21	29.44	29.50	29.56	29.61	29.67	29.72	29.78	29.83	29.89	29.94
22	30.00	30.06	30.11	30.17	30.22	30.28	30.33	30.39	30.44	30.50
23	30.56	30.61	30.67	30.72	30.78	30.83	30.89	30.94	31.00	31.06
24	31.11	31.17	31.22	31.28	31.33	31.39	31.44	31.50	31.56	31.61
−25	−31.67	−31.72	−31.78	−31.83	−31.89	−31.94	−32.00	−32.06	−32.11	−32.17
26	32.22	32.28	32.33	32.39	32.44	32.50	32.56	32.61	32.67	32.72
27	32.78	32.83	32.89	32.94	33.00	33.06	33.11	33.17	33.22	33.28
28	33.33	33.39	33.44	33.50	33.56	33.61	33.67	33.72	33.78	33.83
29	33.89	33.94	34.00	34.06	34.11	34.17	34.22	34.28	34.33	34.39

Fahrenheit	.0°C.	.1°C.	.2°C.	.3°C.	.4°C.	.5°C.	.6°C.	.7°C.	.8°C.	.9°C.
−30	−34.44	−34.50	−34.56	−34.61	−34.67	−34.72	−34.78	−34.83	−34.89	−34.94
31	35.00	35.06	35.11	35.17	35.22	35.28	35.33	35.39	35.44	35.50
32	35.56	35.61	35.67	35.72	35.78	35.83	35.89	35.94	36.00	36.06
33	36.11	36.17	36.22	36.28	36.33	36.39	36.44	36.50	36.56	36.61
34	36.67	36.72	36.78	36.83	36.89	36.94	37.00	37.06	37.11	37.17
−35	−37.22	−37.28	−37.33	−37.39	−37.44	−37.50	−37.56	−37.61	−37.67	−37.72
36	37.78	37.83	37.89	37.94	38.00	38.06	38.11	38.17	38.22	38.28
37	38.33	38.39	38.44	38.50	38.56	38.61	38.67	38.72	38.78	38.83
38	38.89	38.94	39.00	39.06	39.11	39.17	39.22	39.28	39.33	39.39
39	39.44	39.50	39.56	39.61	39.67	39.72	39.78	39.83	39.89	39.94
−40	−40.00	−40.06	−40.11	−40.17	−40.22	−40.28	−40.33	−40.39	−40.44	−40.50
41	40.56	40.61	40.67	40.72	40.78	40.83	40.89	40.94	41.00	41.06
42	41.11	41.17	41.22	41.28	41.33	41.39	41.44	41.50	41.56	41.61
43	41.67	41.72	41.78	41.83	41.89	41.94	42.00	42.06	42.11	42.17
44	42.22	42.28	42.33	42.39	42.44	42.50	42.56	42.61	42.67	42.72
−45	−42.78	−42.83	−42.89	−42.94	−43.00	−43.06	−43.11	−43.17	−43.22	−43.28
46	43.33	43.39	43.44	43.50	43.56	43.61	43.67	43.72	43.78	43.83
47	43.89	43.94	44.00	44.06	44.11	44.17	44.22	44.28	44.33	44.39
48	44.44	44.50	44.56	44.61	44.67	44.72	44.78	44.83	44.89	44.94
49	45.00	45.06	45.11	45.17	45.22	45.28	45.33	45.39	45.44	45.50
−50	−45.56	−45.61	−45.67	−45.72	−45.78	−45.83	−45.89	−45.94	−46.00	−46.06
51	46.11	46.17	46.22	46.28	46.33	46.39	46.44	46.50	46.56	46.61
52	46.67	46.72	46.78	46.83	46.89	46.94	47.00	47.06	47.11	47.17
53	47.22	47.28	47.33	47.39	47.44	47.50	47.56	47.61	47.67	47.72
54	47.78	47.83	47.89	47.94	48.00	48.06	48.11	48.17	48.22	48.28
−55	−48.33	−48.39	−48.44	−48.50	−48.56	−48.61	−48.67	−48.72	−48.78	−48.83
56	48.89	48.94	49.00	49.06	49.11	49.17	49.22	49.28	49.33	49.39
57	49.44	49.50	49.56	49.61	49.67	49.72	49.78	49.83	49.89	49.94
58	50.00	50.06	50.11	50.17	50.22	50.28	50.33	50.39	50.44	50.50
59	50.56	50.61	50.67	50.72	50.78	50.83	50.89	50.94	51.00	51.06
−60	−51.11	−51.17	−51.22	−51.28	−51.33	−51.39	−51.44	−51.50	−51.56	−51.6
61	51.67	51.72	51.78	51.83	51.89	51.94	52.00	52.06	52.11	52.17
62	52.22	52.28	52.33	52.39	52.44	52.50	52.56	52.61	52.67	52.72
63	52.78	52.83	52.89	52.94	53.00	53.06	53.11	53.17	53.22	53.28
64	53.33	53.39	53.44	53.50	53.56	53.61	53.67	53.72	53.78	53.8
−65	−53.89	−53.94	−54.00	−54.06	−54.11	−54.17	−54.22	−54.28	−54.33	−54.3
66	54.44	54.50	54.56	54.61	54.67	54.72	54.78	54.83	54.89	54.9
67	55.00	55.06	55.11	55.17	55.22	55.28	55.33	55.39	55.44	55.5
68	55.56	55.61	55.67	55.72	55.78	55.83	55.89	55.94	56.00	56.0
69	56.11	56.17	56.22	56.28	56.33	56.39	56.44	56.50	56.56	56.6

Fahrenheit	.0°C.	.1°C.	.2°C.	.3°C	.4°C.	.5°C.	.6°C.	.7°C.	.8°C.	.9°C.
−70	−56.67	−56.72	−56.78	−56.83	−56.89	−56.94	−57.00	−57.06	−57.11	−57.17
71	57.22	57.28	57.33	57.39	57.44	57.50	57.56	57.61	57.67	57.72
72	57.78	57.83	57.89	57.94	58.00	58.06	58.11	58.17	58.22	58.28
73	58.33	58.39	58.44	58.50	58.56	58.61	58.67	58.72	58.78	58.83
74	58.89	58.94	59.00	59.06	59.11	59.17	59.22	59.28	59.33	59.39
−75	−59.44	−59.50	−59.56	−59.61	−59.67	−59.72	−59.78	−59.83	−59.89	−59.94
76	60.00	60.06	60.11	60.17	60.22	60.28	60.33	60.39	60.44	60.50
77	60.56	60.61	60.67	60.72	60.78	60.83	60.89	60.94	61.00	61.06
78	61.11	61.17	61.22	61.28	61.33	61.39	61.44	61.50	61.56	61.61
79	61.67	61.72	61.78	61.83	61.89	61.94	62.00	62.06	62.11	62.17
−80	−62.22	−62.28	−62.33	−62.39	−62.44	−62.50	−62.56	−62.61	−62.67	−62.72
81	62.78	62.83	62.89	62.94	63.00	63.06	63.11	63.17	63.22	63.28
82	63.33	63.39	63.44	63.50	63.56	63.61	63.67	63.72	63.78	63.83
83	63.89	63.94	64.00	64.06	64.11	64.17	64.22	64.28	64.33	64.39
84	64.44	64.50	64.56	64.61	64.67	64.72	64.78	64.83	64.89	64.94
−85	−65.00	−65.06	−65.11	−65.17	−65.22	−65.28	−65.33	−65.39	−65.44	−65.50
86	65.56	65.61	65.67	65.72	65.78	65.83	65.89	65.94	66.00	66.06
87	66.11	66.17	66.22	66.28	66.33	66.39	66.44	66.50	66.56	66.61
88	66.67	66.72	66.78	66.83	66.89	66.94	67.00	67.06	67.11	67.17
89	67.22	67.28	67.33	67.39	67.44	67.50	67.56	67.61	67.67	67.72
−90	−67.78	−67.83	−67.89	−67.94	−68.00	−68.06	−68.11	−68.17	−68.22	−68.28
91	68.33	68.39	68.44	68.50	68.56	68.61	68.67	68.72	68.78	68.83
92	68.89	68.94	69.00	69.06	69.11	69.17	69.22	69.28	69.33	69.39
93	69.44	69.50	69.56	69.61	69.67	69.72	69.78	69.83	69.89	69.94
94	70.00	70.06	70.11	70.17	70.22	70.28	70.33	70.39	70.44	70.50
−95	−70.56	−70.61	−70.67	−70.72	−70.78	−70.83	−70.89	−70.94	−71.00	−71.06
96	71.11	71.17	71.22	71.28	71.33	71.39	71.44	71.50	71.56	71.61
97	71.67	71.72	71.78	71.83	71.89	71.94	72.00	72.06	72.11	72.17
98	72.22	72.28	72.33	72.39	72.44	72.50	72.56	72.61	72.67	72.72
99	72.78	72.83	72.89	72.94	73.00	73.06	73.11	73.17	73.22	73.28
−100	−73.33	−73.39	−73.44	−73.50	−73.56	−73.61	−73.67	−73.72	−73.78	−73.83
101	73.89	73.94	74.00	74.06	74.11	74.17	74.22	74.28	74.33	74.39
102	74.44	74.50	74.56	74.61	74.67	74.72	74.78	74.83	74.89	74.94
103	75.00	75.06	75.11	75.17	75.22	75.28	75.33	75.39	75.44	75.50
104	75.56	75.61	75.67	75.72	75.78	75.83	75.89	75.94	76.00	76.06
−105	−76.11	−76.17	−76.22	−76.28	−76.33	−76.39	−76.44	−76.50	−76.56	−76.61
106	76.67	76.72	76.78	76.83	76.89	76.94	77.00	77.06	77.11	77.17
107	77.22	77.28	77.33	77.39	77.44	77.50	77.56	77.61	77.67	77.72
108	77.78	77.83	77.89	77.94	78.00	78.06	78.11	78.17	78.22	78.28
109	78.33	78.39	78.44	78.50	78.56	78.61	78.67	78.72	78.78	78.83

Fahrenheit	.0°C.	.1°C.	.2°C.	.3°C.	.4°C.	.5°C.	.6°C.	.7°C.	.8°C.	.9°C.
−110	−78.89	−78.94	−79.00	−79.06	−79.11	−79.17	−79.22	−79.28	−79.33	−79.39
111	79.44	79.50	79.56	79.61	79.67	79.72	79.78	79.83	79.89	79.94
112	80.00	80.06	80.11	80.17	80.22	80.28	80.33	80.39	80.44	80.50
113	80.56	80.61	80.67	80.72	80.78	80.83	80.89	80.94	81.00	81.06
114	81.11	81.17	81.22	81.28	81.33	81.39	81.44	81.50	81.56	81.61
−115	−81.67	−81.72	−81.78	−81.83	−81.89	−81.94	−82.00	−82.06	−82.11	−82.17
116	82.22	82.28	82.33	82.39	82.44	82.50	82.56	82.61	82.67	82.72
117	82.78	82.83	82.89	82.94	83.00	83.06	83.11	83.17	83.22	83.28
118	83.33	83.39	83.44	83.50	83.56	83.61	83.67	83.72	83.78	83.83
119	83.89	83.94	84.00	84.06	84.11	84.17	84.22	84.28	84.33	84.39
−120	−84.44	−84.50	−84.56	84.61	−84.67	−84.72	−84.78	−84.83	−84.89	−84.94

Centigrade to Fahrenheit

Centigrade	.0°F.	.1°F.	.2°F.	.3°F.	.4°F.	.5°F.	.6°F.	.7°F.	.8°F.	.9°F.
+100	212.00	+212.18	+212.36	+212.54	+212.72	+212.90	+213.08	+213.26	+213.44	+213.62
99	210.20	210.38	210.56	210.74	210.92	211.10	211.28	211.46	211.64	211.82
98	208.40	208.58	208.76	208.94	209.12	209.30	209.48	209.66	209.84	210.02
97	206.60	206.78	206.96	207.14	207.32	207.50	207.68	207.86	208.04	208.22
96	204.80	204.98	205.16	205.34	205.52	205.70	205.88	206.06	206.24	206.42
+95	+203.00	+203.18	+203.36	+203.54	+203.72	+203.90	+204.08	+204.26	+204.44	+204.62
94	201.20	201.38	201.56	201.74	201.92	202.10	202.28	202.46	202.64	202.82
93	199.40	199.58	199.76	199.94	200.12	200.30	200.48	200.66	200.84	201.02
92	197.60	197.78	197.96	198.14	198.32	198.50	198.68	198.86	199.04	199.22
91	195.80	195.98	196.16	196.34	196.52	196.70	196.88	197.06	197.24	197.42
+90	+194.00	+194.18	+194.36	+194.54	+194.72	+194.90	+195.08	+195.26	+195.44	+195.62
89	192.20	192.38	192.56	192.74	192.92	193.10	193.28	193.46	193.64	193.82
88	190.40	190.58	190.76	190.94	191.12	191.30	191.48	191.66	191.84	192.02
87	188.60	188.78	188.96	189.14	189.32	189.50	189.68	189.86	190.04	190.22
86	186.80	186.98	187.16	187.34	187.52	187.70	187.88	188.06	188.24	188.42
+85	+185.00	+185.18	+185.36	+185.54	+185.72	+185.90	+186.08	+186.26	+186.44	+186.62
84	183.20	183.38	183.56	183.74	183.92	184.10	184.28	184.46	184.64	184.82
83	181.40	181.58	181.76	181.94	182.12	182.30	182.48	182.66	182.84	183.02
82	179.60	179.78	178.96	180.14	180.32	180.50	180.68	180.86	181.04	181.22
81	177.80	177.98	178.16	178.34	178.52	178.70	178.88	179.06	179.24	179.42
+80	+176.00	+176.18	+176.36	+176.54	+176.72	+176.90	+177.08	+177.26	+177.44	+177.62
79	174.20	174.38	174.56	174.74	174.92	175.10	175.28	175.46	175.64	175.82
78	172.40	172.58	172.76	172.94	173.12	173.30	173.48	173.66	173.84	174.02
77	170.60	170.78	170.96	171.14	171.32	171.50	171.68	171.86	172.04	172.22
76	168.80	168.98	169.16	169.34	169.52	169.70	169.88	170.06	170.24	170.42

Centigrade	.0°F.	.1°F.	.2°F.	.3°F.	.4°F.	.5°F.	.6°F.	.7°F.	.8°F.	.9°F.
+75	+167.00	+167.18	+167.36	+167.54	+167.72	+167.90	+168.08	+168.26	+168.44	+168.62
74	165.20	165.38	165.56	165.74	165.92	166.10	166.28	166.46	166.64	166.82
73	163.40	163.58	163.76	163.94	164.12	164.30	164.48	164.66	164.84	165.02
72	161.60	161.78	161.96	162.14	162.32	162.50	162.68	162.86	163.04	163.22
71	159.80	159.98	160.16	160.34	160.52	160.70	160.88	161.06	161.24	161.42
+70	+158.00	+158.18	+158.36	+158.54	+158.72	+158.90	+159.08	+159.26	+159.44	+159.62
69	156.20	156.38	156.56	156.74	156.92	157.10	157.28	157.46	157.64	157.82
68	154.40	154.58	154.76	154.94	155.12	155.30	155.48	155.66	155.84	155.02
67	152.60	152.78	152.96	153.14	153.32	153.50	153.68	153.86	154.04	154.22
66	150.80	150.98	151.16	151.34	151.52	151.70	151.88	152.06	152.24	152.42
+65	+149.00	+149.18	+149.36	+149.54	+149.72	+149.90	+150.08	+150.26	+150.44	+150.62
64	147.20	147.38	147.56	147.74	147.92	148.10	148.28	148.46	148.64	148.82
63	145.50	145.58	145.76	145.94	146.12	146.30	146.48	146.66	146.84	147.02
62	143.60	143.78	143.96	144.14	144.32	144.50	144.68	144.86	145.04	145.22
61	141.80	141.98	142.16	142.34	142.52	142.70	142.88	143.06	143.24	143.42
+60	+140.00	+140.18	+140.36	+140.54	+140.72	+140.90	+141.08	+141.26	+141.44	+141.62
59	138.20	138.38	138.56	138.74	138.92	139.10	139.28	139.46	139.64	139.82
58	136.40	136.58	136.76	136.94	137.12	137.30	137.48	137.66	137.84	138.02
57	134.60	134.78	134.96	135.14	135.32	135.50	135.68	135.86	136.04	136.22
56	132.80	132.98	133.16	133.34	133.52	133.70	133.88	134.06	134.24	134.42
+55	+131.00	+131.18	+131.36	+131.54	+131.72	+131.90	+132.08	+132.26	+132.44	+132.62
54	129.20	129.38	129.56	129.74	129.92	130.10	130.28	130.46	130.64	130.82
53	127.40	127.58	127.76	127.94	128.12	128.30	128.48	128.66	128.84	129.02
52	125.60	125.78	125.96	126.14	126.32	126.50	126.68	126.86	127.04	127.22
51	123.80	123.98	124.16	124.34	124.52	124.70	124.88	125.06	125.24	125.42
+50	+122.00	+122.18	+122.36	+122.54	+122.72	+122.90	+123.08	+123.26	+123.44	+123.62
49	120.20	120.38	120.56	120.74	120.92	121.10	121.28	121.46	121.64	121.82
48	118.40	118.58	118.76	118.94	119.12	119.30	119.48	119.66	119.84	120.02
47	116.60	116.78	116.96	117.14	117.32	117.50	117.68	117.86	118.04	118.22
46	114.80	114.98	115.16	115.34	115.52	115.70	115.88	116.06	116.24	116.42
+45	+113.00	+113.18	+113.36	+113.54	+113.72	+113.90	+114.08	+114.26	+114.44	+114.62
44	111.20	111.38	111.56	111.74	111.92	112.10	112.28	112.46	122.64	112.82
43	109.40	109.58	109.76	109.94	110.12	110.30	110.48	110.66	110.84	111.02
42	107.60	107.78	107.96	108.14	108.32	108.50	108.68	108.86	109.04	109.22
41	105.80	105.98	106.16	106.34	106.52	106.70	106.88	107.06	107.24	107.42
+40	+104.00	+104.18	+104.36	+104.54	+104.72	+104.90	+105.08	+105.26	+105.44	+105.62
39	102.20	102.38	102.56	102.74	102.92	103.10	103.28	103.46	103.64	103.82
38	100.40	100.58	100.76	100.94	101.12	101.30	101.48	101.66	101.84	102.02
37	98.60	98.78	98.96	99.14	99.32	99.50	99.68	99.86	100.04	100.22
36	96.80	96.98	97.16	97.34	97.52	97.70	97.88	98.06	98.24	98.42

Centigrade	.0°F.	.1°F.	.2°F.	.3°F.	.4°F.	.5°F.	.6°F.	.7°F.	.8°F.	.9°F.
+35	+95.00	+95.18	+95.36	+95.54	+95.72	+95.90	+96.08	+96.26	+96.44	+96.62
34	93.20	93.38	93.56	93.74	93.92	94.10	94.28	94.46	94.64	94.82
33	91.40	91.58	91.76	91.94	92.12	92.30	92.48	92.66	92.84	93.02
32	89.60	89.78	89.96	90.14	90.32	90.50	90.68	90.86	91.04	91.22
31	87.80	87.98	88.16	88.34	88.52	88.70	88.88	89.06	89.24	89.42
+30	+86.00	+86.18	+86.36	+86.54	+86.72	+86.90	+87.08	+87.26	+87.44	+87.62
29	84.20	84.38	84.56	84.74	84.92	85.10	85.28	85.46	85.64	85.82
28	82.40	82.58	82.76	82.94	83.12	83.30	83.48	83.66	83.84	84.02
27	80.60	80.78	80.96	81.14	81.32	81.50	81.68	81.86	82.04	82.22
26	78.80	78.98	79.16	79.34	79.52	79.70	79.88	80.06	80.24	80.42
+25	+77.00	+77.18	+77.36	+77.54	+77.72	+77.90	+78.08	+78.26	+78.44	+78.62
24	75.20	75.38	75.56	75.74	75.92	76.10	76.28	76.46	76.64	76.82
23	73.40	73.58	73.76	73.94	74.12	74.30	74.48	74.66	74.84	75.02
22	71.60	71.78	71.96	72.14	72.32	72.50	72.68	72.86	73.04	73.22
21	69.80	69.98	70.16	70.34	70.52	70.70	70.88	71.06	71.24	71.42
+20	+68.00	+68.18	+68.36	+68.54	+68.72	+68.90	+69.08	+69.26	+69.44	+69.62
19	66.20	66.38	66.56	66.74	66.92	67.10	67.28	67.46	67.64	67.82
18	64.40	64.58	64.76	64.94	65.12	65.30	65.48	65.66	65.84	66.02
17	62.60	62.78	62.96	63.14	63.32	63.50	63.68	63.86	64.04	64.22
16	60.80	60.98	61.16	61.34	61.52	61.70	61.88	62.06	62.24	62.42
+15	+59.00	+59.18	+59.36	+59.54	+59.72	+59.90	+60.08	+60.26	+60.44	+60.62
14	57.20	57.38	57.56	57.74	57.92	58.10	58.28	58.46	58.64	58.82
13	55.40	55.58	55.76	55.94	56.12	56.30	56.48	56.66	56.84	57.02
12	53.60	53.78	53.96	54.14	54.32	54.50	54.68	54.86	55.04	55.22
11	51.80	51.98	52.16	52.34	52.52	52.70	52.88	53.06	53.24	53.42
+10	+50.00	+50.18	+50.36	+50.54	+50.72	+50.90	+51.08	+51.26	+51.44	+51.6
9	48.20	48.38	48.56	48.74	48.92	49.10	49.28	49.46	49.64	49.8
8	46.40	46.58	46.76	46.94	47.12	47.30	47.48	47.66	47.84	48.0
7	44.60	44.78	44.96	45.14	45.32	45.50	45.68	45.86	46.04	46.2
6	42.80	42.98	43.16	43.34	43.52	43.70	43.88	44.06	44.24	44.4
+5	+41.00	+41.18	+41.36	+41.54	+41.72	+41.90	+42.08	+42.26	+42.44	+42.6
4	39.20	39.38	39.56	39.74	39.92	40.10	40.28	40.46	40.64	40.8
3	37.40	37.58	37.76	37.94	38.12	38.30	38.48	38.66	38.84	39.0
2	35.60	35.78	35.96	36.14	36.32	36.50	36.68	36.86	37.04	37.2
1	33.80	33.98	34.16	34.34	34.52	34.70	34.88	35.06	35.24	35.4
−0	+32.00	+31.82	+31.64	+31.46	+31.28	+31.10	+30.92	+30.74	+30.56	+30.3
1	30.20	30.02	29.84	29.66	29.48	29.30	29.12	28.94	28.76	28.5
2	28.40	28.22	28.04	27.86	27.68	27.50	27.32	27.14	26.96	26.7
3	26.60	26.42	26.24	26.06	25.88	25.70	25.52	25.34	25.16	24.9
4	24.80	24.62	24.44	24.26	24.08	23.90	23.72	23.54	23.36	23.1

Centigrade	.0°F.	.1°F.	.2°F.	.3°F.	.4°F.	.5°F.	.6°F.	.7°F.	.8°F.	.9°F.
−5	+23.00	+22.82	+22.64	+22.46	+22.28	+22.10	+21.92	+21.74	+21.56	+21.38
6	21.20	21.02	20.84	20.66	20.48	20.30	20.12	19.94	19.76	19.58
7	19.40	19.22	19.04	18.86	18.68	18.50	18.32	18.14	17.96	17.78
8	17.60	17.42	17.24	17.06	16.88	16.70	16.52	16.34	16.16	15.98
9	15.80	15.62	15.44	15.26	15.08	14.90	14.72	14.54	14.36	14.18
−10	+14.00	+13.82	+13.64	+13.46	+13.28	+13.10	+12.92	+12.74	+12.56	+12.38
11	12.20	12.02	11.84	11.66	11.48	11.30	11.12	10.94	10.76	10.58
12	10.40	10.22	10.04	9.86	9.68	9.50	9.32	9.14	8.96	8.78
13	8.60	8.42	8.24	8.06	7.88	7.70	7.52	7.34	7.16	6.98
14	6.80	6.62	6.44	6.26	6.08	5.90	5.72	5.54	5.36	5.18
−15	+5.00	+4.82	+4.64	+4.46	+4.28	+4.10	+3.92	+3.74	+3.56	+3.38
16	+3.20	+3.02	+2.84	+2.66	+2.48	+2.30	+2.12	+1.94	+1.76	+1.58
17	+1.40	+1.22	+1.04	+0.86	+0.68	+0.50	+0.32	+0.14	−0.04	−0.22
18	−0.40	−0.58	−0.76	−0.94	−1.12	−1.30	−1.48	−1.66	−1.84	−2.02
19	−2.20	−2.38	−2.56	−2.74	−2.92	−3.10	−3.28	−3.46	−3.64	−3.82
−20	−4.00	−4.18	−4.36	−4.54	−4.72	−4.90	−5.08	−5.26	−5.44	−5.62
21	5.80	5.98	6.16	6.34	6.52	6.70	6.88	7.06	7.24	7.42
22	7.60	7.78	7.96	8.14	8.32	8.50	8.68	8.86	9.04	9.22
23	9.40	9.58	9.76	9.94	10.12	10.30	10.48	10.66	10.84	11.02
24	11.20	11.38	11.56	11.74	11.92	12.10	12.28	12.46	12.64	12.82
−25	−13.00	−13.18	−13.36	−13.54	−13.72	−13.90	−14.08	−14.26	−14.44	−14.62
26	14.80	14.98	15.16	15.34	15.52	15.70	15.88	16.06	16.24	16.42
27	16.60	16.78	16.96	17.14	17.32	17.50	17.68	17.86	18.04	18.22
28	18.40	18.58	18.76	18.94	19.12	19.30	19.48	19.66	19.84	20.02
29	20.20	20.38	20.56	20.74	20.92	21.10	21.28	21.46	21.64	21.82
−30	−22.00	−22.18	−22.36	−22.54	−22.72	−22.90	−23.08	−23.26	−23.44	−23.62
31	23.80	23.98	24.16	24.34	24.52	24.70	24.88	25.06	25.24	25.42
32	25.60	25.78	25.96	26.14	26.32	26.50	26.68	26.86	27.04	27.22
33	27.40	27.58	27.76	27.94	28.12	28.30	28.48	28.66	28.84	29.02
34	29.20	29.38	29.56	29.74	29.92	30.10	30.28	30.46	30.64	30.82
−35	−31.00	−31.18	−31.36	−31.54	−31.72	−31.90	−32.08	−32.26	−32.44	−32.62
36	32.80	32.98	33.16	33.34	33.52	33.70	33.88	34.06	34.24	34.42
37	34.60	34.78	34.96	35.14	35.32	35.50	35.68	35.86	36.04	36.22
38	36.40	36.58	36.76	36.94	37.12	37.30	37.48	37.66	37.84	38.02
39	38.20	38.38	38.56	38.74	38.92	39.10	39.28	39.46	39.64	39.82
−40	−40.00	−40.18	−40.36	−40.54	−40.72	−40.90	−41.08	−41.26	−41.44	−41.62
41	41.80	41.98	42.16	42.34	42.52	42.70	42.88	43.06	43.24	43.42
42	43.60	43.78	43.96	44.14	44.32	44.50	44.68	44.86	45.04	45.22
43	45.40	45.58	45.76	45.94	46.12	46.30	46.48	46.66	46.84	47.02
44	47.20	47.38	47.56	47.74	47.92	48.10	48.28	48.46	48.64	48.82

Centigrade	.0°F.	.1°F.	.2°F.	.3°F.	.4°F.	.5°F.	.6°F.	.7°F.	.8°F.	.9°F.
−45	−49.00	−49.18	−49.36	−49.54	−49.72	−49.90	−50.08	−50.26	−50.44	−50.6
46	50.80	50.98	51.16	51.34	51.52	51.70	51.88	52.06	52.24	52.4
47	52.60	52.78	52.96	53.14	53.32	53.50	53.68	53.86	54.04	54.2
48	54.40	54.58	54.76	54.94	55.12	55.30	55.48	55.66	55.84	56.0
49	56.20	56.38	56.56	56.74	56.92	57.10	57.28	57.46	57.64	57.8
−50	−58.00	−58.18	−58.36	−58.54	−58.72	−58.90	−59.08	−59.26	−59.44	−59.6
51	59.80	59.98	60.16	60.34	60.52	60.70	60.88	61.06	61.24	61.4
52	61.60	61.78	61.96	62.14	62.32	62.50	62.68	62.86	63.04	63.2
53	63.40	63.58	63.76	63.94	64.12	64.30	64.48	64.66	64.84	65.0
54	65.20	65.38	65.56	65.74	65.92	66.10	66.28	66.46	66.64	66.8
−55	−67.00	−67.18	−67.36	−67.54	−67.72	−67.90	−68.08	−68.26	−68.44	−68.6
56	68.80	68.98	69.16	69.34	69.52	69.70	69.88	70.06	70.24	70.4
57	70.60	70.78	70.96	71.14	71.32	71.50	71.68	71.86	72.04	72.2
58	72.40	72.58	72.76	72.94	73.12	73.30	73.48	73.66	73.84	74.0
59	74.20	74.38	74.56	74.74	74.92	75.10	75.28	75.46	75.64	75.8
−60	−76.00	−76.18	−76.36	−76.54	−76.72	−76.90	−77.08	−77.26	−77.44	−77.6
61	77.80	77.98	78.16	78.34	78.52	78.70	78.88	79.06	79.24	79.4
62	79.60	79.78	79.96	80.14	80.32	80.50	80.68	80.86	81.04	81.2
63	81.40	81.58	81.76	81.94	82.12	82.30	82.48	82.66	82.84	83.0
64	83.20	83.38	83.56	83.74	83.92	84.10	84.28	84.46	84.64	84.8
−65	−85.00	−85.18	−85.36	−85.54	−85.72	−85.90	−86.08	−86.26	−86.44	−86.6
66	86.80	86.98	87.16	87.34	87.52	87.70	87.88	88.06	88.24	88.4
67	88.60	88.78	88.96	89.14	89.32	89.50	89.68	89.86	90.04	90.2
68	90.40	90.58	90.76	90.94	91.12	91.30	91.48	91.66	91.84	92.0
69	92.20	92.38	92.56	92.74	92.92	93.10	93.28	93.46	93.64	93.8
−70	−94.00	−94.18	−94.36	−94.54	−94.72	−94.90	−95.08	−95.26	−95.44	−95.6
71	95.80	95.98	96.16	96.34	96.52	96.70	96.88	97.06	97.24	97.4
72	97.60	97.78	97.96	98.14	98.32	98.50	98.68	98.86	99.04	99.2
73	99.40	99.58	99.76	99.94	100.12	100.30	100.48	100.66	100.84	101.0
74	101.20	101.38	101.56	101.74	101.92	102.10	102.28	102.46	102.64	102.8
−75	−103.00	−103.18	−103.36	−103.54	−103.72	−103.90	−104.08	−104.26	−104.44	−104.6
76	104.80	104.98	105.16	105.34	105.52	105.70	105.88	106.06	106.24	106.4
77	106.60	106.78	106.96	107.14	107.32	107.50	107.68	107.86	108.04	108.2
78	108.40	108.58	108.76	108.94	109.12	109.30	109.48	109.66	109.84	110.0
79	110.20	110.38	110.56	110.74	110.92	111.10	111.28	111.46	111.64	111.8
−80	−112.00	−112.18	−112.36	112.54	−112.72	−112.90	−113.08	−113.26	−113.44	−113.6
81	113.80	113.98	114.16	114.34	114.52	114.70	114.88	115.06	115.24	115.4
82	115.60	115.78	115.96	116.14	116.32	116.50	116.68	116.86	117.04	117.2
83	117.40	117.58	117.76	117.94	118.12	118.30	118.48	118.66	118.84	119.0
84	119.20	119.38	119.56	119.74	119.92	120.10	120.28	120.46	120.64	120.8

Centigrade	.0°F.	.1°F.	.2°F.	.3°F.	.4°F.	.5°F.	.6°F.	.7°F.	.8°F.	.9°F.
−85	−121.00	−121.18	−121.36	−121.54	−121.72	−121.90	−122.08	−122.26	−122.44	−122.62
86	122.80	122.98	123.16	123.34	123.52	123.70	123.88	124.06	124.24	124.42
87	124.60	124.78	124.96	125.14	125.32	125.50	125.68	125.86	126.04	126.22
88	126.40	126.58	126.76	126.94	127.12	127.30	127.48	127.66	127.84	128.02
89	128.20	128.38	128.56	128.74	128.92	129.10	129.28	129.46	129.64	129.82
−90	−130.00	−130.18	−130.36	−130.54	−130.72	−130.90	−131.08	−131.26	−131.64	−131.62
91	131.80	131.98	132.16	132.34	132.52	132.70	132.88	133.06	133.24	133.42
92	133.60	133.78	133.96	134.14	134.32	134.50	134.68	134.86	135.04	135.22
93	135.40	135.58	135.76	135.94	136.12	136.30	136.48	136.66	136.84	137.02
94	137.20	137.38	137.56	137.74	137.92	138.10	138.28	138.46	138.64	138.82
−95	−139.00	−139.18	−139.36	−139.54	−139.72	−139.90	−140.08	−140.26	−140.44	−140.62
96	140.80	140.98	141.16	141.34	141.52	141.70	141.88	142.06	142.24	142.42
97	142.60	142.78	142.96	143.14	143.32	143.50	143.68	143.86	144.04	144.22
98	144.40	144.58	144.76	144.94	145.12	145.30	145.48	145.66	145.84	146.02
99	146.20	146.38	146.56	146.74	146.92	147.10	147.28	147.46	147.64	147.82
−100	−148.00	−148.18	−148.36	−148.36	−148.72	−148.90	−149.08	−149.26	−149.44	−149.62

Differences Fahrenheit to Differences Centigrade

Fahrenheit	.0°C.	.1°C.	.2°C.	.3°C.	.4°C.	.5°C.	.6°C.	.7°C.	.8°C.	.9°C.
0	0.00	0.06	0.11	0.17	0.22	0.28	0.33	0.39	0.44	0.50
1	0.56	0.61	0.67	0.72	0.78	0.83	0.89	0.94	1.00	1.06
2	1.11	1.17	1.22	1.28	1.33	1.39	1.44	1.50	1.56	1.61
3	1.67	1.72	1.78	1.83	1.89	1.94	2.00	2.06	2.11	2.17
4	2.22	2.28	2.33	2.39	2.44	2.50	2.56	2.61	2.67	2.72
5	2.78	2.83	2.89	2.94	3.00	3.06	3.11	3.17	3.22	3.28
6	3.33	3.39	3.44	3.50	3.56	3.61	3.67	3.72	3.78	3.83
7	3.89	3.94	4.00	4.06	4.11	4.17	4.22	4.28	4.33	4.39
8	4.44	4.50	4.56	4.61	4.67	4.72	4.78	4.83	4.89	4.94
9	5.00	5.06	5.11	5.17	5.22	5.28	5.33	5.39	5.44	5.50
10	5.56	5.61	5.67	5.72	5.78	5.83	5.89	5.94	6.00	6.06
11	6.11	6.17	6.22	6.28	6.33	6.39	6.44	6.50	6.56	6.61
12	6.67	6.72	6.78	6.83	6.89	6.94	7.00	7.06	7.11	7.17
13	7.22	7.28	7.33	7.39	7.44	7.50	7.56	7.61	7.67	7.72
14	7.78	7.83	7.89	7.94	8.00	8.06	8.11	8.17	8.22	8.28
15	8.33	8.39	8.44	8.50	8.56	8.61	8.67	8.72	8.78	8.83
16	8.89	8.94	9.00	9.06	9.11	9.17	9.22	9.28	9.33	9.39
17	9.44	9.50	9.56	9.61	9.67	9.72	9.78	9.83	9.89	9.94
18	10.00	10.06	10.11	10.17	10.22	10.28	10.33	10.39	10.44	10.50
19	10.50	10.61	10.67	10.72	10.78	10.83	10.89	10.94	11.00	11.06
20	11.11	11.17	11.22	11.28	11.33	11.39	11.44	11.50	11.56	11.61

Differences Centigrade to Differences Fahrenheit

Centigrade	.0°F.	.1°F.	.2°F.	.3°F.	.4°F.	.5°F.	.6°F.	.7°F.	.8°F.	.9°F.
0	0.00	0.18	0.36	0.54	0.72	0.90	1.08	1.26	1.44	1.62
1	1.80	1.98	2.16	2.34	2.52	2.70	2.88	3.06	3.24	3.42
2	3.60	3.78	3.96	4.14	4.32	4.50	4.68	4.86	5.04	5.22
3	5.40	5.58	5.76	5.94	6.12	6.30	6.48	6.66	6.84	7.02
4	7.20	7.38	7.56	7.74	7.92	8.10	8.28	8.46	8.64	8.82
5	9.00	9.18	9.36	9.54	9.72	9.90	10.08	10.26	10.44	10.62
6	10.80	10.98	11.16	11.34	11.52	11.70	11.88	12.06	12.24	12.42
7	12.60	12.78	12.96	13.14	13.32	13.50	13.68	13.86	14.04	14.22
8	14.40	14.58	14.76	14.94	15.12	15.30	15.48	15.66	15.84	16.02
9	16.20	16.38	16.56	16.74	16.92	17.10	17.28	17.46	17.64	17.8

Appendix

Derivatives of Most Common Functions

1. $$\frac{dc}{dx} = 0$$

2. $$\frac{dx^n}{dx} = nx^{n-1}$$

3. $$\frac{du^n}{dx} = nu^{n-1}\left(\frac{du}{dx}\right)$$

4. $$\frac{d(u + v)}{dx} = \frac{du}{dx} + \frac{dv}{dx}$$

5. $$\frac{d(uv)}{dx} = u\frac{dv}{dx} + v\frac{du}{dx}$$

6. $$\frac{d\,\frac{u}{v}}{dx} = \frac{v\frac{du}{dx} - u\frac{dv}{dx}}{v^2}$$

7. $\dfrac{d(\sin u)}{dx} = \cos u \dfrac{du}{dx}$

8. $\dfrac{d(\cos u)}{dx} = -\sin u \dfrac{du}{dx}$

9. $\dfrac{d(\tan u)}{dx} = \sec^2 u \dfrac{du}{dx}$

10. $\dfrac{d(\cot u)}{dx} = -\csc^2 u \dfrac{du}{dx}$

11. $\dfrac{d(\sec u)}{dx} = \sec u \tan u \dfrac{du}{dx}$

12. $\dfrac{d(\csc u)}{dx} = -\csc u \cot u \dfrac{du}{dx}$

Integrals of Most Common Functions

1. $\int u^n \, du = \dfrac{u^{n+1}}{n+1} + C \quad n \neq -1$

2. $\int \dfrac{du}{u} = \ln |u| + C$

3. $\int e^u \, du = e^u + C$

4. $\int \sin u \, du = -\cos u + C$

5. $\int \cos u \, du = \sin u + C$

6. $\int \tan u \, du = -\ln |\cos u| + C$

7. $\int u \, dv = uv - \int v \, du$ (Integration by parts)

8. $\int_a^b f(x) \, dx = F(b) - F(a) \; [F'(x) = f(x)]$ (Definite integral)

9. $\int \cot u \, du = \ln |\sin u| + C$

10. $\int \sec u \, du = \ln |\sec u + \tan u| + C$

11. $\int \csc u \, du = \ln |\csc u - \cot u| + C$

12. $\int \frac{du}{\sqrt{a^2 - u^2}} = \text{Arcsin} \frac{u}{a} + C$

13. $\int \frac{du}{a^2 + u^2} = \frac{1}{a} \text{Arctan} \frac{u}{a} + C$

Letters of the Greek Alphabets Commonly used as Symbols in Various Fields of Science

Α	α	alpha
Β	β	beta
Γ	γ	gamma
Δ	δ	delta
Ε	ϵ	epsilon
Ζ	ζ	zeta
Η	η	eta
Θ	θ	theta
Ι	ι	iota
Κ	κ	kappa
Λ	λ	lambda
Μ	μ	mu
Ν	ν	nu
Ξ	ξ	xi
Ο	o	omicron
Π	π	pi
Ρ	ρ	rho
Σ	σ, ς	sigma
Τ	τ	tau
Υ	υ	upsilon
Φ	φ	phi
Χ	χ	chi
Ψ	ψ	psi
Ω	ω	omega

Index